Chinese Military Modernisation

"If You Know The Enemy And
Know Yourself, You Need
Not Fear The Result Of
A Hundred Battles.

If You Know Yourself But Not
The Enemy, For Your Every
Victory Gained You Will
Also Suffer A Defeat.

If You Know Neither The
Enemy Nor Yourself
You Will Succumb
In Every Battle."

Chinese Military Modernisation

Lt Gen CK Kapur PVSM, AVSM (Retd)

(Established : 1870)

United Service Institution of India (USI)

Field Marshal KM Cariappa Chair

Manas Publications

New Delhi-110 002 (INDIA)

MANAS PUBLICATIONS
(We Convert Fighters into Writers)
(Publishers, Library Suppliers, Importers & Exporters)
4402/5-A, Ansari Road (Opp. HDFC Bank)
Darya Ganj, New Delhi-110 002 (India)
Off: ✆ 011-23260783, 23265523, Res: 011-23842660
Fax: 011-23272766
E-mail: manaspublications@gmail.com
manaspublications@yahoo.com
Website: www.manaspublications.in

Third Impression **2014**

ISBN: 13 Digit - 978-81-7049-159-0
10 Digit - 81-7049-159-2

₹ 595/-

Designed at
Manas Publications

Printed in India at
RK Print Services, Delhi
and published by Mrs Suman Lata for
Manas Publications, 4402/5-A,
Ansari Road, (Opp HDFC Bank),
Darya Ganj, New Delhi-110 002 (INDIA)

Contents

Preface

The Chinese Armed Forces traditionally hold pivotal position in the society, particularly the Government. No study on China can be complete without a proper understanding of its Armed Forces. Encouraged by the spectacular economic growth, China is vigorously pursuing numerous programmes for the modernisation of its armed forces. It is modernising conventional and nuclear forces to create a limited regional power projection capability in the near future. This is to supplement its foreign policy objectives; close the immense technological gaps with advanced Western powers and strive to improve nuclear strike capability in its pursuit to be a great power.

This book attempts to examine the salient features of modernisation of the Chinese Armed Forces, its military status and its implications on India's national security with a view to formulate a comprehensive and long term strategic response. Authentic information about China is scarce; about its Armed Forces, even more scarce.

Compartmentalised comparisons on certain economic-socio-military aspects between these two countries that are basically so different tend to be deceptive. However, some comparisons have been attempted in order to emphasise important issues relating to India's national security. An effort has been made to cover the gaps through personal interaction with a large number of Indian and foreign Sinologists, who have lived and worked in China over a number of years. I am indebted to them for their help. My personal gratitude to the Directorate General of Military Intelligence, Army Headquarters for making available various unclassified publications and documents which provided valuable insight in the functioning of the People's Liberation Army. My special thanks to Lieutenant General Satish Nambiar, PVSM, AVSM, VrC (Retd) and Major General Y K Gera (Retd) and other staff members of the United Service Institution of India, who rendered all possible help for successful completion of this book.

Lt Gen *C K Kapur* PVSM AVSM, (Retd.)

Chapter 1

Introduction

China has been and continues to remain an enigma to many of us. Because of its size, history, culture, ideology and system of governance, much that happens inside China is not often known to outsiders or even to many inside the country. China's sheer size and growing power are altering the contours of Asian security, international trade and global balance of power. In fact China's emergence as a major power constitutes one of the most significant strategic events of the present day. The uncertainties about China's future capabilities and intentions have bedevilled analysts and scholars in general in the academia, strategic and political 'think tanks', corporate circles and the media worldwide.

China is the largest nation in Asia, and the most populous country in the world. Its land boundaries with 14 countries measures over 28,000 km. It is bordered by Russia to the north; Mongolia to the north-west, Korea to the north-east, CIS, the erstwhile part of USSR, Afghanistan and Pakistan to the west, the East and South China Seas and the Pacific to the east and Vietnam, Myanmar, India, Bhutan and Nepal to the south. It has a coastline of 14,500 km, having maritime boundaries with North Korea and South Korea, Japan (Senkaku Island), Taiwan, Philippines, Malaysia, Brunei and Vietnam in view of its claim over Spratly and Paracel Islands. It has an Exclusive Economic Zone (EEZ) of 1.3 million sq km. China has an estimated population of 1.2 billion comprising 56 ethnic nationalities with an annual growth rate of 1.2

percent. China's urban population is about 18 per cent of the total, one-third of which is concentrated in 21 mega cities. Thus attacking China's cities, either in a nuclear strike or in a punitive attack, does less damage to China than to any other power. To defeat China, the core must be invaded and occupied, and demography renders both the tasks very difficult. Approximately 94 percent of China's population is Han, while the remaining 6 percent comprise the minorities. The Han can be found predominantly in the Yellow, Yangtze and Pearl river valleys and the plains of north east. The ethnic minorities, though small in numbers (6 percent), inhabit 64 percent of the total area. They reside mainly in Xinjiang, Tibet, Ningxia and Guangxi. However, it was not until the 12th Century that Southern China was effectively conquered by Han migration and the South is now included in what is called 'Core China' (45 percent of territory). Manchuria was included in the 19th Century and the grip on Inner Mongolia, Xinjiang and Tibet is even more tenuous. While the 'outer' empire of Mongolia, Tibet and Xinjiang comprised 55 percent of modern day China, it is home to only 10 percent of the population. The presence of minorities in the border areas is a source of sensitivity, insecurity and instability, though it poses no challenge to 'Core China', which has a Han majority. Unlike India, this partially explains the basic political and ideological continuity of China over the last fifty years despite numerous upheavals.

The geographical isolation and its size granted China a degree of immunity unknown in Europe and many other parts of Asia. The Zhou dynasty claimed China from Yangtse to the Yellow River and beyond around 1100 BC. However, it was in 300 BC that a unified and legalistic state of Han China was formed. During Ming period (1368-1644) China became a major maritime power while a fragmented India's Chola Dynasty in the south undertook limited naval exploits in Far and Middle East. The Chinese civilisation for many centuries remained unaware of other centres that were equal to its level of achievements and it thus considered itself as the sole civilisation in the world. Even the Chinese term for its country is 'Zhong Guo', which means Middle Kingdom, around which the whole universe is revolving. According to this Sino-centric theory, national boundaries were not acknowledged, because the societies of neighbouring states were considered extensions of China itself. This phenomenon of Sino-centrism has had a profound and enduring impact on Chinese world outlook, which has persisted till present times. The traditional Chinese feeling of superiority over the 'foreign devils' (a term synonymous for non-Chinese) received a severe setback when Western penetration in the 19th Century resulted in a series of "Unequal Treaties" (commencing with the Treaty of Nanjing in 1842), which limited the sovereignty of China, forcing her to open up her ports to foreign trade and cede her territory to "Western Treaty Powers". The British attempted to make it a land of poppy eaters, crushed the 'Boxer Rebellion' against opium trade and imposed heavy reparations. This domination of a "people as proud as their past glory and civilization as the Chinese" by the 'foreign devils', which lasted for nearly a century, was a humiliation that has conditioned the thinking of all Chinese leaders and has proved a stimulus to them to restore China to a position of eminence in the international arena.

In September 1931, Japan attacked China and, within a day, occupied many towns in Southern Manchuria. When final deliverance came in August 1945, China had truly become independent after a 100 years. The price it paid was colossal – approximately 1.2 million soldiers, two million civilians and $49 billion in economy. The China which stood up and looked around, was a China seething with anger and vengeance, determined to exact revenge for 100 years of suffering and humiliation. It was not prepared to let anything stand in its way. After numerous sacrifices, over a period of time, the Chinese learnt how to protect their national integrity and self respect without depending on any help from others. China's main goal at present is to erase, or at least compensate for, that memory. China is trying to gain the respect of the world, even the awesome respect of the great powers. The most important and effective tool to achieve liberation was the People's Liberation Army, the PLA, which encompasses its Army, Navy, Air Force, Police and the para-military forces.

Strategic Culture

China's deep-rooted traditional strategic culture needs to be taken note of. In the period prior to Alexander's ambitious plans to conquer the world, a Chinese military philosopher, Sun Tzu, wrote such prophetic principles of military warfare that they have guided Chinese and other leaders to the present day. Sun Tzu's *The Art of War* has as much relevance today as it had almost 2,500 years back. Sun Tzu's precepts were studied with great interest not only by the Chinese but by the Japanese, the Russian, the German, the British and the American military thinkers. These were extensively applied during the Second World War and in campaigns thereafter. There are striking similarities between the modern Maoist model and classical Chinese military traditions. For example, "The army is the fish and the people are the water" is first found in the works of Mencius, a disciple of Confucius who wrote in the Third Century B.C. The principle of isolating and surrounding an enemy to destroy him is basic to the ancient Chinese game of 'weiqi,' better known in the West under its Japanese name 'go.' The Great Wall of China, an incredible feat of engineering, was extended largely in 214 BC. Besides being one of the seven wonders of the world, it is reputedly the only man-made object visible from the moon, stretching as it does for about 7000 km. It was created against forays of barbaric Turks and Mongols. This happened during the period when India was repeatedly ravaged by invaders from the north. Though economically prosperous, the India of that time was fragmented and failed to make any viable attempt to defend itself. In the 15th Century, Chinese Navy headed by the famous eunuch Admiral Zheng, carried out several important expeditions. During 1407-33, he visited 29 Asian and African countries. During Ming period in the 17th Century, China was a major maritime power. The Chinese leaders have traditionally displayed a deep understanding of global balance of power, geo-strategy, *real politik* and diplomacy. *Ab initio,* Chinese leaders were very clear about their long-term goals and objectives. Even before China became independent officially, Mao secretly sent his personal emissary Field Marshal Liu Shaoqi to Moscow in early 1949, to reach an understanding with Stalin for sharing of atomic secrets with China. The Chinese emissary made it clear that Chairman Mao would only visit the Soviet Union after China became independent, if the Soviets were prepared to assist China in developing nuclear capability. China embarked on an indigenous nuclear programme soon after 'Liberation' in October 1949. It utilised primitive rudimentary organisations for this programme in 1954; strategic missiles programme in 1956; nuclear submarines and submarine launched ballistic missile programmes in 1958. Interestingly, the Chinese leaders began to pursue such advanced strategic systems programmes without waiting for the proper development of indigenous scientific base or infrastructure. The leaders thus set the pace for the technological growth of the armed forces. In the early 1950s, China moved forces into Tibet to protect its under-belly. The Indian leaders hardly understood the long-term strategic ramifications of this move. Though highly unprepared, China confronted the American military might during the Korean War, to keep American forces away from their borders. The reversion of Hong Kong and Macau to China during late 1990s are success stories of Chinese long-term strategy and diplomacy.

During its formative years, China was a revisionist power. It militarily confronted the USA (1950s), India (1962), the erstwhile Soviet Union (1967) and Vietnam (1979). In early Sixties, it severed ties with its giant Communist neigbour the USSR on ideological grounds when the latter refused to share key nuclear secrets. This resulted in over 10,000 Russian scientists, technicians and advisors quitting China, leaving its fledgling nuclear programme in the lurch. China has not hesitated to use force or coercion to press for its maritime claims in South China Sea. In late Nineties, it became a major arms exporter to developing countries. A dynamic, pragmatic and flexible approach has all along paved the way for China's march towards higher international status. Since its progressive opening up in the 1990s and its constructive engagement with the USA, some Indian analysts have opined that China has evolved into a rising status quo power. Perhaps this thesis may have to be put through another decade of post-Cold War uncertainties for it to be objectively validated.

The centrality of the PLA in the socio-political fabric of China has been evident in the seven decades since its establishment in 1927. Notable benchmarks have been the liberation struggle during 1930s-40s, the liberation of Tibet (1950s), the Great Cultural Revolution (1960s), crackdown on student demonstrators at Tiananmen (1989), and the installation of President and Party Chairman, Jiang Zemin at the 'core' of China's Third Generation Leadership (1990s). The recent elevation of Vice Premier Hu Jintao as the Vice Chairman of the all powerful Central Military Commission and the fact that he was being spoken of as the likely successor of President Jiang Zemin, further supports the continuing symbiotic relationship between the Party and the Gun.

India and China: Symmetries Versus Asymmetries

The historical similarities between China and India include their self-image as dominant centres of civilisation. For China, this ancient cultural superiority was recognised in the bordering states of Xinjiang, Tibet, Mongolia, Manchuria, Korea, Japan, and Vietnam; for India, it was recognised in the Southeast Asian periphery of Myanmar, Thailand, Cambodia, and Indonesia. The asymmetries of development of India and China, rather than the similarities, are indeed noteworthy. China's consolidation of a unified state as early as 221 BC and the continuous attempts to protect the heartland by controlling the periphery, added a military dimension to China's security strategies, that relied on naturalistic and symbolic tributary exchanges. By contrast, India remained a civilisational construct, rather than a unified state, until well into the 19th Century. The subcontinent was divided into local kingdoms, ruled by warring kings who placed a higher value on symbolic and ritual forms of superiority rather than on military control of conquered territory. This made possible the conquest of India by the Turko-Afghans from the 10th Century, and eventually the British rule.

China's strategic culture is a study in contrasts vis-à-vis India. During the Indian independence struggle, the leadership, with the exception of Netaji Subhash Chandra Bose, had no exposure to military matters. Such was their lofty image for the India of their dreams that the military was even considered unnecessary by some. A vast majority of our political thinkers and intelligentsia betrayed total disregard for military history, geo-strategic realities, the lessons of European colonial wars and the two World Wars. In mitigation, it may

Chapter 2

The Socio-Economic Developments

The dynamic growth of the Chinese economy over the last 50 years ranks among the most spectacular developments of the 20th Century. However, this development has been the result of many drastic changes in its economic policies from state run "communes" to "market economy with Chinese characteristics" under the leadership of Mao Zedong, Deng Xiaoping and the regime of Jiang Zemin. However, the basic economic policies can be readily distinguished from that of Deng Xiaoping (1978-95). Jiang, being a protege of Deng, generally continued in the footsteps of his master, fine-tuning his policies in the initial stages of his rule from 1996 onwards. The Chinese have their own traditional peculiar ways, be it external relations, internal issues or economic matters. The major socio-economic changes were often initiated by some symbolic brief utterances of their charismatic leaders. Mao's statement that "communes are good" was enough to ensure large scale communising of the whole country-side. Deng's slogan, "to get rich is glorious", triggered the expansion of Special Economic Zone in the south and accelerated the pace of economic reforms.

Economic Policy

Mao took a major decision of adopting the Soviet-type economy. Most of the economic institutions were nationalised and placed under state control. The basic strategy of economic growth was not demand driven but investment and supply driven, resulting in rapid restructuring and growth of the economy in the initial stages. The Chinese brought in considerable degree of decentralisation and their system was fairly loose and more dependent on local cadres' initiatives and management skills than the erstwhile Soviet Union. There were some adaptations of the Soviet type economic model which proved counter productive thus bringing Mao's economic policy to an end. A great emphasis was laid on self dependency during the initial economic development period. This policy limited China's participation in global economy just at the time when global technological development and economic growth were generating tremendous benefits to the developing economies pursuing the export promotion policy. Indian economy also suffered because of our initial close door policy. Further, to ensure people's participation, the reliance on mobilisation campaign and mass movements to achieve the rapid transition to true socialism resulted in devastation of the economic as well as the social growth of China during that period. In 1956 Mao initiated "the Hundred Flowers Campaign" by asking the elitists and intellectuals to "bloom and content" by expressing their criticisms, but when the response was not as `flattering' as the Chairman expected, the intellectuals were suddenly punished with the Anti-Rightist campaign of 1957 which sent thousands of educated Chinese to rural labour, and paralysed China's intellectual life for more than two decades or so. The other leading examples were the Great Leap Forward (1958-59) and the Cultural Revolution (1966-69) resulting in decline in the output, and the famine that followed is estimated to have cost the Chinese 10 years of positive growth. By 1976, with the death of Mao and arrest of his most radical followers soon after, Mao's economic policy regime came to an end.

Despite some major shortcomings, the Chinese economy during 1952-75 did exhibit some important accomplishments. China's gross domestic product (GDP) grew at an average annual rate of 6.7 percent and the industries increased its share of GDP from 20.9 to 45.7 percent. In addition they achieved a significant industrialisation of Chinese economy with remarkable degree of self sufficiency. During the Mao era, Chinese did provide for widespread educational opportunities, better access to medical and health care, more opportunities for women, a significant social welfare package for the urban workers – some positive aspects which were related to the Soviet type of economy and were followed for overall social and economic upliftment of the masses. To that extent it is fair to say that Mao succeeded to a large extent to put China on some sound footing both socially and economically despite certain failings.

Mao's death in 1976 and arrest of the "Gang of Four" shortly thereafter provided an opportunity for a change in both leadership and the overall economic development strategy. While some within the leadership wanted to continue Mao's policy, the 3rd Plenum of the 11th Chinese Communist Party (CCP) Central Committee in November-December 1978 extended full support to Deng Xiaoping for modifications of the economic policies which resulted in great economic strides for the country in the years ahead. Deng obviously focused on eliminating most serious economic problems inherited from Mao. Agriculture was the most pressing of these problems and local cadres and peasants were allowed to abandon the communes and restore household farming. This was a major shift from the past. Deng's policies opened the Chinese economy to foreign trade and investment.

Furthermore, non-state enterprises, both domestic and foreign, were created to compete with the state owned enterprises. The development of market economy with non-state enterprises saw a dominant share of transactions and profits made by the units outside the direct control of the government. Deng thus initiated a new process of building the Chinese economy based on various global developments. However, this process of building a "socialist economy with Chinese characteristics" was far from complete when Deng died in 1997. During Deng's regime, the growth rate of GDP was 9.8 percent or approximately 1.5 times the average annual growth rate under Mao.

The change in economic policies in the Deng period involved a significant decentralisation of decision making throughout the economy and withdrawal of state interference in the normal economic activities of the people. By 1996, China was the world's seventh largest participant in the world trade and direct foreign investment had gone to approximately 3 lac enterprises in China, the foreigners accounting for 40 percent of US $ 650 billion invested in those enterprises. This radically changed the pattern of livelihood of the average Chinese. Whereas the per capita household consumption expenditure (in constant prices) was growing by 2.2 percent a year during the Mao era, the annual rate of growth increased to 7.3 percent under Deng.

The Economic Policy of Jiang Zemin

Jiang neither had the stature nor the experience which his predecessors had enjoyed because of their being a part of struggle for independence as well as their linkage with the PLA. He, therefore, chose to fine tune the earlier policies that were initiated by Deng for the creation of a stable "socialist market economy with Chinese characteristics". Jiang, therefore, proceeded on the path laid by his mentor to complete his unfinished task of a self sustaining Chinese economic growth and social stability. In the process of overall economic development China as a nation has set a target for itself of attaining the economic and technological level of a developed country by 2050 i.e. within 100 years of the Communist party coming to power.

Economic Record Card

China's economic development in the last two decades has been phenomenal. Chinese society has experienced some unprecedented drastic socio-economic changes. While it took USA forty seven years to double its per capita income, it took China only nine; in fact China has quadrupled its per capita income in the last 18 years. It took the USA 50 years to reduce the share of the labour force engaged in agriculture from 70 percent to 50 percent. China accomplished the same change in 17 years. The World Bank in its study, "China 2000: Development Challenges in the New Century", claims that, "China has telescoped into one generation what other countries took centuries to achieve. No country has tried to accomplish so much in such short a time. China's unique attempt to complete two transitions at once – from a command to a market economy and from a rural to an urban society–is without a historical precedent." China has averaged, in diverse political conditions more than nine percent real economic growth ever since its reforms began in 1978. During the period 1992-95 it has maintained a growth rate of 12-13 percent. From just US $ 70 billion in 1985, China's overall foreign trade had grown to US $ 235 billion by 1994. In 1995, China's foreign trade accounted for four percent of the world's merchandise export and 3.4 percent of the world's merchandise imports. By 1998, the China was one of the ten largest trading nations in the world, up from thirty-first in 1980. The World Bank had predicted that China is set to become the world's largest economy in 2010. Using the purchasing power parity calculation of Gross Domestic Production (GDP), China already had the world's second largest economy (4.45 trillion USD) in early 1999 next only to the USA (8.88 trillion USD).

To get a first hand impression about the country, China's performance parameters in socio-economic field pertaining to a few key areas are given in Table-1. The corresponding figures for India have also been shown to facilitate a comparison.

Table 1

SOME SOCIO-ECONOMIC COMPARISONS

		CHINA	INDIA
**	Literacy Rate (%)	81.5	52.1
**	Population Growth (%) £	0.87	2.1
**	Life Expectancy £	71.86	62.0
**	GDP (Trillion) (Purchasing Power Parity)	5.56	- -
**	GDP Growth (%)	7.3	5.8
*	Per Capita Income (Yearly)	1000	450
**	Population Below Poverty Line (%)	10	27
	Savings Rate (%)	40	22.3
*	Foreign Exchange Reserves (Bn) £	286.4	70.3
*	FDI Inflow (Bn) in 2001	46.8	2.3
*	Trade Balance (Bn)	524.2	3.8

Note: All figures are for 2000-2001, unless otherwise stated.

* US Dollars

Sources: ** CIA – The World Fact Book 2002-portion on China; *China Statistical Yearbook*; *Indian Economic Survey*;

£ Figures for 2002. *The Economic Times*, New Delhi, 13 January 2003.

Chinese Economy–Some Vulnerabilities

Despite having a commendable track record in socio-economic development so far, it will be too simplistic to assume that everything is well with the Chinese economy, taking into account its size, population, diversities and aspirations. The transaction from a planned to a market economy has also produced many upheavals which have made China a land of contradictions and gross inequalities. While its strengths are well documented, we should have an idea of its vulnerabilities to facilitate objective evaluation of a possible direction that China is heading for in the next two decades or so.

Chinese leadership pushed the economic reforms and consequently the economic growth much further and faster that anyone thought feasible. But to-day that "Euphoria" seems to have petered out. With the negative gravitational pull of some 200 million urban and rural unemployed, and a sharp decline in exports and foreign investments, Chinese leadership has been forced to reverse or put on hold major economic reform policies. The decline in the growth rate of GDP during the last few years has been as follows :

Year	Growth Rate (%)
1992	14.2
1993	13.5
1994	12.6
1995	10.5
1996	9.6
1997	8.8
1999	8.0
2000	8.0
2001	7.1*
2002	7.5*

* *Source:* People's Daily - http://english.peopledaily.com.cn/200201/04/eng 20020104-88033.shtml

In Shanghai the commercial buildings completed in 1997-98, the occupancy rate is below 20 percent. The State's biggest economic conundrum is the State-Owned Enterprises (SOE), which account for nearly 50 percent of China's industrial output and employ at least 50 million workers, yet two-third of the nearly 15000 large and medium sized production units are unable to meet the wage-bill of its workers on a regular basis. Keeping the SOEs afloat severely strains State coffers. Urban un-employment is growing and due to its fear of adding millions to the unemployment rolls, the government has repeatedly postponed reforms. The intractability of a large number of inefficient SOEs and a government controlled banking system, more debt-ridden than any other in Asia, are strong indicators of feeble health of Chinese economy in years ahead, notwithstanding the rampant corruption and worsening regional and class-wise distribution of incomes. In short, the Chinese leadership must seek out some new engines of growth to sustain the erstwhile momentum enjoyed under Deng, which will not be easy to find.

There has been a substantial decline in average per capita availability of farmland, from about 0.18 hectare (ha) in 1949 to just below 0.08 ha in 1995. If these figures are accurate, China's land availability would rate in the same class as Bangladesh, and the only other populous nations with less arable land would be Egypt, Japan and Korea. Interestingly the European average is nearly 0.3 ha and the North American mean is almost 0.7 ha As arable land is shrinking, China's foodgrain production is also declining. With a fifth of the world's population, China has only a fifteenth of its arable land. Today China is a net importer of foodgrains (approximately 15 million tons in 1995) which may reach around fifty million tons in the early part of this century. By 2030 it could reach ninety million tons or roughly half of total current global grain exports.

More than 200 of China's large cities experience chronic housing, water and electricity shortages. China today has an estimated 120 million illegal rural migrants flowing from the countryside to the over-crowded cities. In Guangzhou alone, more than a million migrants –a sixth of entire population — lack residence. China is a net oil importer. Chinese consume only 40 percent per capita energy of the world average. They use relatively large amount of coal, which meets 75 percent of primary energy demands. With five times the population of the US and ten times that of Japan, China has huge, latent potential demand for energy that will manifest as economy develops and consumerism catches up further.

Balancing Economic Liberalisation and Political Freedom

How will China balance its economic liberalisation with its dictatorial Communistic single party system of governance and limited personal freedom? It is also striving to be a major military power having regional domination with global reach. Economic liberalisation and its globalisation will urge its people to seek greater personal freedom that is a shift towards democratic system of governance. There are nine political parties in China, the Chinese Communist Party and eight 'democratic parties'. These 'democratic parties' are in no sense opposition parties but they do participate in political dialogue at national level–a consultative approach. However, these parties really have no choice but to accept the CCP's leadership. They survive on sufferance from the CCP and could be immediately disbanded if they ignore the CCP's bidding. How is it likely to cope up with these contradictory requirements? Is it likely to go the Russian way? Is the last bastion of Communism likely to crumble under the weight of its economic globalisation? Such a contingency appears to be unlikely. During the last 50 years, Chinese have demonstrated a great economic and political pragmatism. They have their own way of steering through, of innovating and creating what may be termed a way out with typical Chinese characteristics, as they are capable of living with a host of contradictions. They have done it in the past and there is no reason as to why they should not be able to do so in future. The practice of crystal gazing China politically, economically and strategically through a Westerner's prism needs correction. "Things," as China watcher Lucien Pye says, "are never as black or white, as good or as bad, as Westerners tend to expect them to be".

An Assessment

In early Fifties, both India and China were industrially and agriculturally backward and followed the Russian model with considerable variations for their economic growth notwithstanding their different types of political systems. Over a period of time three major variants placed China in an advantageous position. Firstly, through one-child norm and its ruthless implementation, China has been able to control its population better than India. Soon India will become the most populous country in the world. Consequently, China achieved major successes in the spread of education, health care and poverty alleviation programmes. Secondly, China laid greater emphasis on the development of basic infrastructure than India thus setting a stable base for its economic take off. In the field of transportation and communication, a great headway has been made in railways, highways, ports, shipping and civil aviation including 80 international air routes to about 50 cities. China today has an installed capacity of more than three lakh megawatts of electricity as against India's 90,000 MW. China produces 100 million tonnes of steel while we are still less than 30 million. Thirdly, China has displayed remarkable pragmaticism in implementing a smooth transition from a state-controlled economy to a market one. Political stability, continuity in top leadership, disciplined workforce and the role of Non-Resident Chinese–a rich business class that had spread all over the world–substantially contributed towards its all round development. What better proof is there of its well rounded development than the fact that even in sports China stands next to the United States with a tally of 80 medals in the first Olympic of this century; India had to be satisfied with one bronze. In sum, it is fair to conclude that though China is long way off from becoming a real global economic power, despite all internal dissentions and contradictions, it will continue to make a steady economic progress at par or even better than the general trends in global economic growth. It will, therefore, be able to generate sufficient funds to support the various programmes for the modernisation of its defence forces. Our neighbour in the north is well poised to be economically stable and militarily strong–a reality which we cannot afford to ignore.

Chapter 3

Defence Modernisation
The Prelude

The Chinese leadership moved in a systematic, deliberate and planned manner towards the modernisation of its armed forces over a period of time. The Chinese armed forces, in fact the Army, grew from rag-tag rural based guerrilla bands with highly decentralised, diffused command and control structure. Being a continental power, its armed forces mainly consisted of the Army. The Navy and the Air Force came up as subordinate support arms. In 1949, the Red Army as it was then called, propelled the Chinese Communist Party to power after 28 years of battling Kuomintang, Japanese and warlord forces. Just as the PLA was consolidating power, it was severely tested on the battlefield in Korea. Despite sustaining huge losses (an estimated three million), Mao's "volunteers" held their own against the United Nations forces. But the experience convinced Chinese commanders of their backwardness and the pressing need for modernisation. Mao had justified a thorough programme of modernisation for the Armed Forces in 1954 on grounds of China's status as a "great power". "Our industry, agriculture and military are insufficient", Mao had remarked. "Imperialists assess you in terms of these things and therefore bully us. They say, `do you have the atomic bomb?"

Since the formation of the independent Communist China, its first 25 years were of socio-economic turmoil and political upheaval, next 20 years for the revival of economy and the period thereafter is being utilised to make it militarily a global player backed up by a viable and self sustaining economy. Economically, China has taken giant strides in that direction. Military modernisation was justifiably placed last amongst its 'Four Modernisations' at national level. It must be understood that modernization of the PLA is an ongoing process with varying emphasis from time to time. For the Middle Kingdom, national pride comes naturally to the Chinese. It is not only a "sleeping giant" waking up but giant rising with an inherent sense of wounded pride and resultant annoyance that it had been battered and cheated by the rest of the world. Hence it is natural that it takes comprehensive all round measures for the growth of its most vital organ which is solely responsible for its birth and largely responsible for its stability and acceptability, both internally as well as globally. The recent economic growth has only strengthened the process of modernisation of the Chinese Armed Forces.

Political Aims

There is deep linkage between the strategic culture and strategic capabilities of a nation. China is striving hard to carve a rightful place for itself in the world, though it is fully aware that this is a task which is not easy to achieve and will take considerable time to come about. China's perceived political aims in the foreseeable future are as follows :

(a) To achieve the status of a developed country by 2050 AD.

(b) Strive for technological and economic parity with other developed nations to the maximum possible extent.

(c) Maintain territorial integrity and ensure continued Communist Party rule.

(d) Play dominant role in all global affairs and project itself as a major 'pole' in the developing multi-polar world order.

(e) Attempt to re-unify Taiwan and other disputed or claimed territories.

(f) Exercise tacit control and domination over its periphery.

In order to achieve the aforesaid objectives, China considered it prudent to bring about organisational and structural changes, which would assist in the modernisation of its armed forces. The overall framework for functioning of the modernised PLA was set on a firm footing by the 1982 Constitution of the People's Republic of China. Article 29 emphatically stated:

> *"The armed forces of the People's Republic of China belong to the people. Their tasks are to strengthen national defence, resist aggression, defend the motherland, safeguard the people's labour, participate in national reconstruction and work hard to serve the people.*
>
> *The State strengthens the revolutionization, modernization and regularization of the armed forces in order to increase the national defence capability."*

Special measures were taken to define the laws and procedures governing the PLA and other organs, apart from other regulations, as can be surmised from the PRC's avowed aims for "revolutionization, modernization and regularization."

Lessons from Korean, Vietnam and Gulf Wars

In the Korean War, despite highly outdated weaponry, China succeeded in achieving its strategic aim of keeping the Americans at bay from its borders. The Chinese thereafter planned for the modernisation of its armed forces at an affordable pace because of non-availability of funds. The dismal performance of the Indian Army during its war with

China in 1962 brought in some complacency in its planned modernisation programmes. In its war with Vietnam in 1979, while trying "to teach them a lesson" they learnt an important lesson themselves, that its Army and its weapons and equipment even in low level conventional warfare was not up to the mark and needs to be urgently upgraded. However, the Chinese underplayed the importance of this war since it was not a typical modern war. It was referred to as a "special war within a special war" being limited in scope and the absence of its tri-service character. However, the PLA's poor performance was taken note of and it rekindled the process of modernisation. Subsequently Deng Xiaoping as Central Military Commission Chairman, talked about the necessity of reforming the PLA's structure for preparing the armed forces for "an early war, a major war and a nuclear war".

Not many countries have studied the Gulf War in such great detail as China. To assess the lessons of Gulf War, China ordered a number of studies. The Gulf War acted as a catalyst in the overall modernisation process. The Chinese realised that they are at least two to three generations behind the USA and they need to take some immediate steps to narrow the gap. It taught Beijing some crucial lessons:

(a) That the prospects of a global conflict have indeed diminished, and the prospects of regional and local conflicts have increased.

(b) That a weak state would remain vulnerable to pressure, intimidation and ultimately to aggression.

(c) That conventional military power, still the predominant component of 'comprehensive national strength' in the emerging New World Order, is a function of superior technology.

(d) Electronic warfare was decisive to the result of the entire war.

(e) Air and Naval powers were critical arms in modern warfare.

(f) Overall capability is measured by rapid response and fast deployment.

(g) Logistics support capability is as important as actual fighting strength.

STRUCTURAL AND ORGANISATIONAL REFORMS

In order to enhance the overall operational effectiveness of its armed forces the Chinese leadership introduced many wide-ranging structural and organisational changes during the early Nineties to lay down a sound base for the absorption and progress of its modernisation programmes, akin to preparing the soil before sowing the seed for a good crop. From the higher defence organisation to recruitment procedures many changes were introduced at various levels, new laws were passed, several existing departments were amalgamated and command and control system was streamlined to make it more responsive to meet the likely challenges ahead. Manpower was reduced to make the Armed Forces more technically oriented and less manpower intensive as has been the case in the past. Special emphasis was laid on education and training of troops and senior commanders. Some of the noteworthy changes are discussed in the succeeding paras.

Central Military Commission (CMC)

The Central Military Commission (CMC) replaced the People's Revolutionary Military Commission and in terms of Article 93 of the Constitution of the PRC (1982), it is responsible for evolving the over-all military strategy of the nation and "directing the armed forces of the country". It coordinates issues involving PLA with the other organs of the

State including defence industrial growth, R&D, budgeting and training. In 1984 it was placed under the direct supervision and control of the Politburo of the Chinese Communist Party (CCP). CMC is involved in all major aspects of defence and acts as the single "command organisation providing unified leadership for the armed forces". In a pyramid-like structure of the Chinese military establishment, the CMC acts as the highest decision making body of the PLA. Over the years, the heterogeneous nature of the CMC has not been conducive for the professional and scientific growth of military ethos, traditions and culture. There were numerous political military factions competing for eminence within the CMC. Its membership over the years has been reduced as follows:

Year	Members
1977	66
1987	9
1997	7
2001	5*

* One Chairman, three Vice Chairmen. *Source*: Special Report, China's New CMC Lineup. www.vic.info.org/RegionsTops.ns

Over a period of time there have been continuous attempts to enhance the civilian control over the PLA. The PLA membership in the Politburo has now been reduced from 50 percent during the Cultural Revolution to the present level of two out of 21 members. There is, at the political level, a trend which is likely to get strengthened as the old generation of revolutionary military leadership fades away. However, at the functional level, the PLA general departments enjoy considerable autonomy. The administrative and operational control exercised by the Ministry of National Defence is minimal unlike in India where the Ministry of Defence enjoys vast administrative and financial control over the armed forces and virtually functions as a Supreme Headquarters without any accountability.

Re-organisation of Departments and Headquarters

The bloated size of various headquarters has been considerably reduced and their powers curtailed to professionalise the system of chain of command and control. The headquarters of major arms like Artillery, Armour, Infantry and Engineers were abolished in early Eighties. Correspondingly some functional arms departments were created within the General Staff Department, much smaller in size and primarily dealing with doctrinal, training and administrative coordination. There were similar reductions at the Military Region levels. All this not only resulted in reduction of manpower but also enhanced the authority of the PLA and its four major departments.

There were some important reasons to bring the modernisation of the PLA within the ambit of constitutional framework. The chaos of Cultural Revolution heavily undermined the overall prestige of the PLA and adversely affected its discipline and military cohesion. The organisational reforms, which took place as a prelude to the serious modernisation efforts, brought in such changes which would not only eliminate all perceived procedural obstructions but would smoothen and hasten the overall modernisation process. The first casualty was the Culture Department which was one of the sub-departments of the General Political Department. Traditionally, the Culture Department looks after arts and letters within the PLA. It was this department which was instrumental in initiating the Cultural Revolution within the PLA. The surgical severance of the Cultural Department and its Liaison Office was a clear message that unlike in the past, the PLA will tread the professional path rather than getting embroiled in internal party politicking.

In the field of Defence Research and Development, a single unifying institution, Commission of Science and Technology and Industry for National Defence (COSTIND) was formed by amalgamating three central institutions i.e. CMC Science and Equipment Commission, PLA National Defence Science and Technology Commission and the National Defence Industry Office of the State Council. This resulted in laying greater emphasis on Defence R&D and its interface with civilian sectors of the economy. The PLA General Armament Department (GAD) was established as the fourth general department in April 1998 for better management of production and procurement of weapons and equipment. GAD is the single most important institution for the induction of hi-tech weapons, R&D in military hardware and the coordination of military with civilian sectors of economy.

Re-organisation of Military Regions (MRs)

The Military Regions (MRs) occupy an important place in Chinese military hierarchy. These are akin to theatre commands or war zones with responsibility for conduct of operations within their area of jurisdiction with full command and control of the forces operating in that region. Traditionally the Commanders and the Political Commissars (PCs) wield considerable influence not only within the PLA but also in the overall administration of the provinces within their regions. During the Cultural Revolution, the MRs virtually became independent with varied areas of interest – a trait not conducive to professionalism and unity of command by the four major departments of PLA. To imbibe the spirit of modern warfare some of the following changes were brought in over a period of time.

(a) In 1955 the number of MRs was increased to 13 from 12. This number was reduced to 11 in 1971. Again in 1985 MRs were reorganised and their number was reduced to seven. As a result of these reforms their operational boundaries were re-drawn depending on the nature of threat perception.

(b) The staff at MR headquarters was reduced by 40 to 50 percent during the reforms in 1985. Similarly the authority of MR on appointments, promotions and dismissals as well as on move and relocation of military units was drastically curtailed. These powers—administrative and operational—were passed on to the State Council or to PLA Headquarters. This enhanced the central authority of the PLA over its command theatres and field armies.

In the restructuring of field formations, 36 Field Armies were reorganised into 24 Group Armies with a pronounced emphasis on joint tri-service operations. A similar reorganisation was done in the Air Force and Navy resulting in having seven Air Districts and three Fleet Commands in addition to the Strategic Missile Force or the Second Artillery as it is popularly known. Most of these reforms were based on the deep analysis of the US "Air-Land Battle" concept and combined arms operations. The Group Armies were also suitably organised to be more responsive to the requirements of modern warfare under hi-tech conditions. Tank divisions were done away with. Instead integrated mechanised armoured divisions now form part of reorganised Group Armies.

Notwithstanding the aforesaid, the next round of reorganisation is already under discussion. It is believed that China is likely to reorganise its military regions as integrated zonal commands for better absorption of 'Combined arms operations' in joint land-sea-air warfare.

Manpower Rationalisation

Over the last two decades there has been a considerable reduction and rationalisation in the PLA manpower. PLA's biggest strength during its long struggle for independence was its large manpower which became one of the impediments to its high-tech modernisation programmes. The PLA manpower restructuring though initially planned for in the early Eighties met with stiff opposition from 'old guards' within the Party. It was after its dismal performance in its war with Vietnam that the Party High Command decided that a 'lean and mean' standing armed forces would be a necessity if its modernisation programmes are to go through. The over-all strength of the PLA during the last two decades has been as follows :

Year	PLA Manpower (in millions)
1980	4.2
1987	3.23
1992	2.9
1997	2.7
2000	2.5
2002	2.27

Thus over a period of time approximately 2 million Service personnel (including civilians) are believed to have been demobilised. Almost 81 percent of the overall reduction has been in the ground forces while 19 percent reduction has taken place in the Air Force and the Navy. More than 30 units and formations at or above Corps level were disbanded and over 300,000 officers were compulsorily retired or transferred to civilian posts. In a major rationalisation of manpower, the PLA Capital Construction Corps was institutionally transferred to civilian control and the PLA Railway Corps was transferred to the Ministry of Rail-Roads during the years 1982-85. There is no change in their basic role. These are being extensively used in Tibet for roads and rail construction but not directly under PLA, consequently showing reduction in its strength. The total manpower thus lost was approximately 510,000.

While undoubtedly certain major manpower reductions have taken place, but to believe that the PLA manpower has shrunk by approximately two million may not be correct, when one looks at the overall increase in the strength of the People's Armed Police (PAP) (1.2 million strong) and over five million reservists who form a part of armed forces as a second line of defence. The total forces under arms still remain nearly nine million. China thus continues to have the largest armed forces in the world. In essence the exercise in downsizing the PLA is based more on rationalisation and restructuring than in outright reduction.

Reintroduction of Ranks

One of the most important reforms was the reintroduction of military ranks. The system of officers' ranks was introduced in 1955. Considered as symbol of elitism in the Army of 'Comrades' these were abolished just before the onset of Cultural Revolution when the PLA returned to its revolutionary ideals of a classless society and a military force without ranks. There were several reports in 1978 that the ranks were being restored but it was only in 1988 that it actually happened. A new officers' service law was enacted in 1988 which specified conditions of service and rules for promotion. The mandatory proportion of generals to other officers was specified. The levels at which officers' appointment and promotion was to be made were re-examined. The powers of Military Regional Commanders in this respect were curtailed with greater centralisation at CMC. The post of Assistant Chief of General Staff existed till 1984.

From 1984 to 1992, this post was abolished, and was revived in November 1992 by the 14th Party Congress. Many new laws and regulations were notified between 1992 and 1996 symbolising the rejection of extreme egalitarian ethos of the past and a renewed commitment of making PLA a modern Army. However it was in 1997 that the Chinese enacted a National Defence law.

In 1999, conscription service period was reduced to two years for non-technical services and three years for technical services and Second Artillery. Thus the proportion of conscripts in the PLA from 82 percent of its total strength will come down to less than 65 percent by 2003. However, the age profile of an average PLA soldier will be maintained at about 20 years. Thus some far reaching changes regarding its manpower policies were introduced to bring the PLA at par with trends in other armed forces across the globe.

Regularisation

The Chinese term for regularisation refers to a wide range of related issues. These include main areas of standardisation, training, officer education, doctrine, the conduct of military operations and the establishment of legal norms and regulations that regulate these activities, as well as military discipline and so forth. Regulations provide guidelines as to how training, education, promotion and combat operations are to be carried out. Discipline codes and regulations create universal standards for punishment, rewards and administrative procedures. All of these elements are necessary for the successful conduct of modern military operations.

For Chinese leaders specifically, regularisation of the country's armed forces has two main goals. First and most important, it is supposed to strengthen and codify the leadership of the CCP over the armed forces. Second, it is intended to help create proper conditions for Defence modernisation. The system of officers' ranks was abolished just a year before the onset of the Cultural Revolution. The chaos of the Cultural Revolution heavily undermined discipline and military cohesion. Regulations provide order, discipline and cohesiveness to the armed forces; they define the mission and role of the military within society.

A new military service law was enacted in 1984, as were regulations on discipline, punishments, rewards, and internal service matters. A new conscription law was enacted in 1985. The rate of creation of new laws and regulations increased between 1992 and 1996. In 1992, the General Political Department set up judicial administration departments to unify and manage judicial administration work within the PLA. The focus on NCOs is significant as they provide the skill and experience crucial to the effective functioning of any military. By focussing on NCOs, the revised military service law is clearly intended to improve the PLA by building up a strong contingent of career NCOs capable of carrying out a variety of military tasks.

Doctrinal Reforms

Doctrine essentially is a set of guidelines on planning and the use of force in pursuit of military objectives. It is perhaps best thought of as an outline for what type of armed forces a country wants to build and what sort of mission those armed forces are to carry out. Doctrine helps shape and mandate what types of organisations should exist within the armed forces. China's economic development since 1979 has put greater emphasis on the need to defend the increasingly well developed coastal areas and protect trade routes. Not only are there separate career tracks for military officers and commissars, it is also now the norm to serve in the same service arm for their entire career. In the 1970s it was not unusual to see a political commissar transfer from the infantry to the Navy. The most striking aspect of the past 20 years of regularisation, however, is how long the process has taken and how recent most of the developments have been.

Implementing Rule of Law in PLA

Serious questions remain as to whether the new regulations prohibiting PLA involvement in commercial activities promulgated in 1999 will be fully implemented. While there is some debate over how much the PLA as a whole benefited from these enterprises, there is no doubt that many individuals and units profited at the expense of their warfighting capabilities. PLA enterprises actually encouraged military units and individual officers to lie to superiors and hide illegal activities. The problem of officers not thinking of their behaviour as being governed by legal restrictions persisting was amply demonstrated by the involvement of military officials in the Yuanhua corruption scandal in Xiamen. This smuggling case is reported to have involved as much as $10 billion, making it the biggest corruption scandal in the history of the PRC. In addition to civilian and military officials in Xiamen, the scandal has also been reported to involve some very senior national level military officials. Another factor that may have an impact on the growth of a rule of law in the PLA is that an increasing number of officers are being exposed to Western military thought and norms. About 20,000 PLA officers have been sent abroad for military study during the period 1992-1997, and many countries have sent active duty and retired military personnel to China to lecture and teach. The PLA may finally be becoming institutionalised; the open question that remains is whether or not it will be too little, too late. A clear answer is yet to emerge.

Defence Expenditure

Military expenditure is the monetary form of military policy and a comprehensive expression of the country's capabilities and intentions. Countries pursue different military policies and defence expenditure is based on economic strength and determined by the assessment of domestic and international, political and military situation.

China wanted to ensure a sound economic base before embarking on a large scale comprehensive modernisation of its armed forces. The vibrating Chinese economy ensured continuous flow of sufficient funds for this purpose. In specific areas, China has been allocating considerable funds even during the period when its economy was not even at the take off stage. For example, in the light of the January 1955 decision on nuclear weapons, "funds for science and technology" rose from about US$15 million in 1955 to about US$100 million in 1956; the Chinese Academy of Sciences received three times as much money in 1957 as it had received in 1956.

China is a closed society. It is difficult to get hard information on military matters. The extent and outlays of Chinese military expenditure is largely unknown. If some of the official Chinese reports are to be believed their defence budget has had a negative growth. The Chinese also argue that their military expenditure is only $6 per capita compared to $1,100 of the US, thus being only two hundredth of the US expenditure. The flaw in relying solely on official Chinese defence budget figures is that hardly anyone believes that they fully account for the real military expenditures. State released figures for PLA spending are thought to cover only the costs of personnel (military pay and some fringe benefits) and of operations and maintenance (O&M). The Ministry of Nuclear Industry is producing nuclear weapons; the Ministry of Ordnance Industry, arms and ammunitions; the Ministry of

Aeronautics Industry is building military planes; and the Ministry of Transport deals with matters relating to ships. Areas believed to be wholly or partially missing from the Chinese defence budget are :

Research and Development : Funding for military R&D is covered under other departmental heads. The cost of China's nuclear forces are hidden within the nuclear energy and space programmes.

Procurement : PLA procurement is absent from official military budgets.

People's Armed Police (PAP) and PLA Reserves : Funded from provincial budgets.

Military Pensions : Paid by the Ministry of Civil Affairs.

Agriculture Production of PLA owned Farms : The produce helps feed the troops and is also sold in the market. The annual earnings are believed to be around $1 billion.

Earning from Overseas Arms Sales : In early Nineties China exported nearly $ 2-3 billion worth of arms each year which go directly to the PLA.

Earnings from PLA's Non-military activities : Some estimates put the earning as high as $20 billion per year, at least during the period when PLA involvement in commercial ventures was legally permitted.

The official Chinese defence expenditure figure, released in 1979 for the first time, at $ 12.9 billion was 18 percent of the planned government expenditure. Even at that time the International Institute of Strategic Studies (IISS) estimated the Chinese defence budget to be nearly $ 50 billion. The official Chinese defence budget in 1997 was only $ 9.7 billion which even a casual observer would find unbelievable. The allocation for Defence in 1998 is $ 11.1 billion (increase of 12.9 percent). For the period 2001-2002, Beijing announced a 17.7 percent increase in defence spending, the biggest expansion in real terms in the last two decades. It needs to be noted that by all estimates during last 20 years or so, while India's Defence expenditure remained constant except some yearly spurts in-between, China's Defence expenditure has risen consistently, both overtly and covertly. The PLA has not been cash starved at least during the last 10 years or so. During visits of foreign military observers, Chinese commanders hide their capabilities to gain technological assistance and to surprise future enemies. China's unexpected breakthrough in its nuclear weapons programme also illustrates this strategic trait. The Chinese are sensitive to such analysis. In no case do they allude to expensive hi-tech modernisation being undertaken by the PLA.

It is assessed by sources in Taiwan and Japan that the actual Defence expenditure in China is three to eight times the official Defence budget.

As per the *SIPRI Military Expenditure Database,* India's military expenditure at constant 1998 prices and exchange rates has varied from 9390 million US $ in 1998, 10733 in 1999, 11793 in 2000 and 12879 in 2001. As a share (%) of the gross domestic product it has varied from 2.2 in 1998, 2.4 in 1999 and 2.4 in 2000. As far as China is concerned, military expenditure varied from 19000 million US $ in 1998, 21100 in 1999, 23100 in 2000 and 27000 in 2001. As a share (%) of the gross domestic product it has varied from 2.0 in 1998, 2.1 in 1999, and 2.1 in 2000.

Overhaul of Training Philosophy

The older generation of officers and soldiers though having considerable operational experience, were generally ill educated and consequently unable to grasp the nuances of the present day hi-tech warfare. Increasing the educational standard of troops and overhauling the training philosophy is another important reform intended to turn the PLA into a modern fighting machine. The emphasis has been to make it more technology intensive rather than manpower intensive as has been the case in the past. Most of the other ranks had hardly any formal education, while less than 50 percent of the officers had education beyond senior middle school. Now the minimum educational standard for other ranks is junior high school (10th class) while all officers will be graduates at the time of recruitment or formally imparted education upto that level during the pre-commissioning training. The PLA has also introduced a Reserve Officers Training Corps system. For military schools, there are five levels of degree programmes. These programmes are dispersed in about 600 specialties. At present there are 21 authorised PLA institutions and 138 specified programmes for doctorate degrees and over 500 for post-graduation degrees. This has encouraged PLA all ranks to acquire a higher level of civil education. In 1978 only eight percent officers had exposure to education at college level, while in 1999, this proportion is approximately 71 percent. Some 85 percent of the Divisional Commanders hold masters or higher degrees.

China issued its revised General Training Programme in 1996 with emphasis on Joint Operations, opposing force training and live fire exercises. A full-fledged divisional level force along with support elements has been earmarked to act as 'enemy troops' in all major exercises, a system copied by PLA from the erstwhile USSR. A major innovation is the Field Army Combat Training Establishment where integrated armies are trained. The PLA today has 100 training institutes as against 40 a decade ago. As many as 21 universities covering 138 disciplines offer upto Ph.D level courses for PLA personnel. More than 50 percent technical officers are either scientists or engineers. Further an increasing number of PLA officers are being exposed to Western military thought and norms, at a level and scale which has never been experienced in the past.

National Defence University. China set up its National Defence University (NDU) in 1985. National security has very wide ranging and comprehensive ramifications much beyond the bounds of territorial integrity and sovereignty – ranging from food, water, energy, communications to environmental security. All these and many other aspects need to be dealt with through an inter-disciplinary approach for evolving a balanced and all inclusive concept of security. In China, NDU was formed by merging the colleges of the three General Departments as a part of modernisation drive to have the nation's highest military institution of learning to train and equip the high-ranking officers to meet the challenges of future warfare. It is the supreme institution of military training in China. It functions directly under CMC and is presided over by a very senior and experienced high ranking service officer. Its main tasks are:

(a) To train operational, political and logistics officers at and above Army levels, and the equivalent in the Air Force and the Navy, senior staff for the Headquarters at and above the MD levels, senior research fellows and senior officials from government agencies.

(b) To conduct research into strategic issues and questions related to defence modernisation.

(c) To provide advice and consultancy to the CMC and the General Headquarters of the PLA.

(d) To study, evolve and revise contemporary operational doctrines and concepts based on 'global geo-strategic developments'.

Organisationally, the NDU is divided into four departments–Training Department, Political Department, Scientific Research Department and Administrative Department. The 12 teaching divisions work on strategy, Marxist theory, operational art of war, CCP history, command

and control, international economy and politics, overseas logistics and foreign languages. Specific departments are devoted to strategic studies, Marxist theoretical studies and armed forces development studies. The NDU curriculum has about 100 subjects and in the last 15 years since the founding of the NDU, a total of 400 courses were offered at different periods and about 1500 textbooks were compiled and revised among which more than 100 were listed as "key projects". In the recent past, it has received approximately 5000 foreign experts, defence personnel and others from about 80 countries.

The NDU has created a degree programme and educational system for graduate students. By 1997, 244 students obtained their master's degrees. To facilitate the development of multi-dimensional talents for military purposes, the NDU recruited doctoral candidates and by 1999, twenty of them had obtained their doctorate. In reality China's NDU is playing a more important role in the military as well as diplomatic fields than ordinarily visible to the outside world. In September 2000, it hosted the Fourth ASEAN Regional Forum Meeting of the Heads of Defence Colleges at Beijing in which 21 member-countries including India participated. Topical subjects like Asia-Pacific security at the turn of the century, role of ASEAN in the 21st Century and Information Warfare were discussed. It is an important institution giving valuable inputs for defence and foreign policy formulation. Even in the USA, the NDU functions as a high level think-tank at the disposal of the Defence Secretary and Joint Chiefs of Staff. Its importance is obvious from the fact that President Bush chose to announce the National Missile Defence policy from its podium than elsewhere. Though its necessity is well appreciated, India is yet to have its national defence university whose scope and span of activities will be much wider than the existing National

Defence College. The setting up of the national defence university is a requirement which we cannot afford to delay any longer.

Political Training : The General Political Department (GPD) is the oldest and vital organ of the PLA through which the Party controls the 'gun'. Originally this function was performed by General Cadre Department but in July 1958 it got amalgamated with GPD. The GPD functions through a hierarchy that parallels the military chain of command. This remains one of the crucial differences from most of the other professional modern armed forces in the world. The GPD chain of command is a pyramid of Party committees going through every military unit upto a battalion level to inject the Party's ideological work throughout the PLA organs emphasising discipline, morale and motivational aspects. In essence this system provides enough leverage to Party to pursue its policies in the PLA. The Red Army regulations of 1930-1932, in 34 parts, forms the basis of this system which largely remains unchanged till now.

The arrangement of the Political Department in the organisational hierarchy is exhaustive and elaborate. All army units above the regimental level have Party committees, battalions, primary Party committees, and companies, Party branches. The Party committee or Party branch forms the core of leadership in each army unit. Units above the regimental level and independent battalions have political commissars; battalions and companies have political instructors. These political cadres take care of the routine matters of the Party committees and Party branches. They and the military commanders are jointly in charge of the work of their units. In addition, there are political departments in the units above the regimental level, which manage the work of the Party and other political work under the leadership of the Party committees and the political commissars. At each level, the political department has an equal rank of importance with the administrative headquarters. The political committees are tasked to implement the political line, measures and policies of the CCP and national laws and edicts among the army personnel. After Tiananmen incident when some army units disobeyed orders to open fire on their countrymen, the emphasis on imparting ideological education has increased. Maintaining internal stability and studying President Jiang Zemin's teachings are amongst 10 top ideological tasks the PLA officers and soldiers have been asked to pursue in the year 2001-2002. The national and regional army authorities have circulated a list of areas in which indoctrination is to be boosted amongst the rank and file. Priority areas include boosting party "leadership over the gun", raising people's awareness of national defence, ensuring that the PLA remained a "guarantee" for national reunification and boosting its role in maintaining internal stability and security. Other priorities stressed the politics or orthodox Marxist values, ensuring that more "paragon figures" emerged from the ranks, developing the tradition of "plain living and hard struggle", learning how to win in hi-tech warfare, studying Jiang Zemin's teachings on military modernisation and taking part in the "anti-imperialist struggle" in the world arena. Some leaders appear worried that officers and men would soon succumb to "quasi-capitalist poison" in the new market economy. The revival of plain living and hard struggle is aimed at restoring Maoist values of frugality and total obedience to authority.

Conclusion

The need to modernise its armed forces has been well understood and appreciated by the Chinese leadership particularly after the Korean War. However, due to lack of industrial and technological infrastructure and paucity of funds they could hardly make any worthwhile progress. The economic boom in the Eighties changed the situation drastically. It was only after the Gulf War that the Chinese grasped the impact of emerging science and technology on modern warfare and seriously embarked on a modernisation plan for their armed forces – a selective, phased and time bound plan of action to acquire some specific military capabilities. The approach was multidirectional and multidimensional with 'Chinese Characteristics'. Prior to embarking on specific programmes, they set the stage for a smooth take off for their various modernisation programmes in conformity with their short, medium and long term geo-strategic goals.

Chapter 4

Changing Security Doctrines

China has had no major threat since the demise of the erstwhile Soviet Union and the collapse of Russia's economy, yet it retains one of the world's largest armour and mechanized force structures. Much of the PLA's experimentation since 1992 has focused on a large scale ground fight against a sophisticated opponent. Yet the likelihood of the US Army invading China, or of anyone else being able to even contemplate such an attack, seemed remote even to the PLA. China already possesses an enormous military advantage over its neighbours, at least for a conflict on its periphery within easy reach of its Military Region-based logistical network. Training to confront a high-tech opponent does make China's ground forces better, but without a ground threat it does not increase China's useable military capability.

In China, military doctrine is determined at the highest level of political and military leadership. Military doctrine reflects the Chinese Communist Party's views on global strategy, nature of war and the Armed Forces follow the Party line in preparing themselves for various contingencies.

China's strategic thinking and planning underwent a drastic revision under the 'Third Generation' Chinese leadership with President/Chairman Jiang Zemin at its core. A major element of this new thought process has been to evolve a new strategy for peacetime Defence modernisation as an integral part of China's national construction. China's opening up has necessitated awareness about operational war-fighting doctrines. 'Seeking Peace' has been projected as the central part of China's declared national security policy.

Limited High Technology War

Since its liberation in October 1949, China's conventional military doctrines initially evolved from 'People's War' to 'People's War Under Modern Conditions'. Late Chairman Mao and his comrades developed this doctrine during the Long March and Liberation Struggle. With the demise of Mao, the second-generation leadership under Deng evolved the doctrine of 'Limited/Local War' during the late 1970s, which was modified by 'Modern Conditions' by end of 1980s. After the Gulf War, the concept of 'Local Wars Under High Technology Conditions' has become the doctrine in vogue. Chinese analysts at the National Defence University, Beijing, have theorised that China could face five types of modern limited war contingencies. These are border conflicts, contention for territorial seas and islands, surprise air attacks, resistance against partial intrusions, and punitive counter attacks against neighbouring states. The new doctrine of 'Limited War Under high-technology Conditions' is still a paper concept which has not yet been worked out in detail. Till about 2020 or so, China is unlikely to have the hardware for a high-tech war. However, Chinese military thinking conceptually has dwelt on the following main characteristics of modern limited wars:

(a) The strategic objectives are limited and initial phases of the war are characterised by decisive combat actions.

(b) High technology, low intensity, complete replacement of earlier time and space factors and development of multi-dimensional battlefields.

(c) Increased speed of military operations, the need for quick decisions and increased importance of surprise factor.

(d) Multi-dimensional combined arms warfare with well trained/equipped, small and compact contingents being the main force in battle groups.

(e) Due to the larger scale of warfare, the organisation and command set-ups are more complex.

(f) In the high-tech modern warfare scenario, Chinese have upgraded the importance of logistics than ever before. The logistics support will determine its operational reach beyond its borders, which is likely to remain short for quite sometime to come.

(g) Diplomacy will be active during the war to keep it limited.

Local Wars

The Chinese strategists opined that though there is no possibility of any large scale conventional or nuclear war, medium and small-scale wars will be difficult to avoid. However, 'Local War' is an inappropriate translation of Chinese term 'jubu zhanzheng', which seems to include a broad range of scenarios, encompassing, in fact, almost any war smaller in scale than a global war or a major nuclear war.

The threats to China's security were analysed in detail by China's National Defence University, Academy of Military Sciences and other 'Think Tanks' prior to the commencement of various modernisation programmes. Based on the review of the global security environment and the situation prevailing on China's periphery, there has been a general consensus amongst Chinese military thinkers that the local wars pose the greatest threat to China's security. This view is based on numerous destabilising factors that prevail on China's land and maritime boundaries and a large number of territorial disputes which China has with its neighbours. The Chinese military thinkers firmly believe that historically China's borders have been unstable and the neighbouring countries have taken advantage of China's weakness from time to time. India, Vietnam and Japan figure prominently in Chinese strategic calculus in relation to the territorial challenges on its borders. A perusal of some of the articles written by the PLA officers on Sino-Indian 1962 War reveal that the Chinese 'counter attack' in 'self-defence' was fully justified to strike at India's 'swollen arrogance'. Notwithstanding the historical evidence to the contrary, China believes that India is in possession of 1,25,000 sq. km. of its 'resource rich' territory which is thrice the size of Taiwan. According to the PLA, the area which is under India's illegal control is not only strategically important but is also rich in hydel power with its reserves having approximately 30 percent of the country's entire hydel power resources. It is further alleged by China that 90,000 sq.km. area in India's Eastern sector lies mostly in the tropical region and is important for the development of Tibetan economy. This area can be further developed to meet the agricultural requirements of Tibet, which is primarily dependent on the mainland. Many of the Chinese military thinkers have opined that to oppose 'Indian expansionism' and to solve the India-China border problem it is necessary for China to perceive India as an 'active enemy'.

Whilst such views have been rebutted by India expectedly and are not very much in focus presently, these could find favour with China's hierarchy if it perceives that the USA is exploiting India to contain China. It is noteworthy that the recent Indo-US bonhomie and even the non-commercial ambiguous Indian support to the Bush administration over National Missile Defence (NMD) is viewed with concern in China.

In January 1999, Central Military Commission Chairman Jiang Zemin signed an order implementing the first major revision to PLA operational doctrine since the mid-1980s. The old regulations, referred to as the "first-generation" effort to standardise PLA operational procedures, had become outdated. They did not address joint service operations, or warfare under high-technology conditions. They also did not cover the full range from national and campaign-level down to unit-level operations, and obviously did not incorporate the lessons of the Gulf War. The evolving concept of Revolution in Military Affairs was not addressed. The 13 new combat regulations, known collectively as the "Chinese People's Liberation Army Joint Campaign Programme," are designed to rectify these omissions.

Active Defence

This is an old concept which has been upgraded and modernised over a period of time. Mao had defined 'Active Defence' as a doctrine of 'offensive defence' or a concept of deterrence through decisive engagement which prescribed strategic defence and tactical offensive. Like other aspects of Chinese strategic thought, Active Defence has also been given new interpretations. The old Maoist concepts like 'protraction, strategic retreat' for 'luring the enemy deep inside' are no longer considered workable. China's new production centres and industrial infrastructure being vulnerable, the emphasis has shifted towards 'forward defence' or even 'positional offensive'. Among the various kinds of wars envisioned in the Active Defence doctrine, the Chinese do not visualise a world war or a long term war against China happening in the next 50 years or so. However, a border conflict or a limited war is a distinct possibility. This is the basic thought around which China's Active Defence policy has been reframed. It is based on the premise that in the short or even medium term, China must prepare for conflicts which would be limited, intense and far more frequent due to the growing struggle of the regional powers around China's land or maritime boundaries. Such engagements will have limited strategic objectives necessitating rapid responses for quick decisions. Active Defence is therefore offensive in nature. At tactical level, a pre-emptive strike is a prime expression of 'Active Defence'. The Chinese consider this compatible with the principle of "Striking Only After the Enemy has Struck". A National Defence University publication has clearly mentioned that "in China's anti-aggression wars in the future, all military activities conducted by the enemy and aimed at breaking up China territorially and violating its sovereignty *de facto* constitute the first shot in strategic terms. The actual combat that follows is merely a continuation and expansion on the strategic level of the first shot. For this reason, when a point is reached at which the enemy is amassing troops or engaged in other military activities, targeting this nation and war has become inevitable, we must lose no time in launching a pre-emptive attack which is an important example of active defence strategic thinking at the start of a limited high tech war".

Nuclear Doctrine

Even prior to its liberation, the Chinese leadership always understood the importance of nuclear weapons in the game of global power politics. Over a period of time the Chinese leaders generally agreed that in addition to buying self esteem and status, nuclear weapons have a military utility as well. On 16 October 1964, China exploded the atom bomb, on 27 October 1966, it conducted the first successful missile test and on 17 June 1967, it exploded the first hydrogen bomb. In 1958, the Special Artillery Corps was raised. On 1 July 1966, the Second Artillery Corps was officially established with the approval of CMC. The Second Artillery Corps today is a well trained strategic missile – nuclear and conventional–force with a high level of strike capability. The Chinese have declared time and again that China was compelled to develop nuclear weapons and its purpose for doing so was to break the nuclear monopoly. It also wished to oppose nuclear blackmail, to defend its independence and security and last but not the least to safeguard world peace. The Chinese analysts predicted that membership of the nuclear club would increase to 23, including Brazil, Iraq, Libya, Japan, Vietnam and South Korea, by the year 2010. Therefore, if China is to play an important role in global affairs, like the other permanent members of the UN Security Council, it should have a well pronounced nuclear capability. Before China formally ratified the Comprehensive Test Ban Treaty in 1996, it rushed through a series of nuclear tests that were considered essential for the miniaturisation of warheads, especially for submarine-launched missiles. On 21 May 1992, China had its largest ever underground nuclear test with a yield of one megaton. This was 70 times the size of the bomb dropped on Hiroshima in 1945, and considerably exceeded the United States self imposed test limit of 150 kt.

Limited deterrence now is the basic Chinese nuclear doctrine which implies having enough capabilities to deter conventional, theatre and strategic nuclear wars and to control and suppress escalation during the nuclear war, thereby responding to any kind or any level of tactical and strategic nuclear attack from any source. Chinese were quite clear in their mind that implementing such a nuclear strategy will require the following :

(a) A greater number of smaller, more accurate, survivable and penetrable intercontinental ballistic missiles (ICBMs)

(b) Tactical and theatre nuclear weapons to hit battlefield and theatre military targets and to suppress escalation

(c) Ballistic missile defence to improve the survivability of the limited deterrent, as well as to defend key military command centres and national facilities

(d) Space-based early warning and command-and-control systems; and

(e) Anti-satellite weapons to hit enemy military satellites.

During the last three decades or so, in its transition from "hiding" to "fighting", there has been a perceptible change in the PLA's attitude towards nuclear weapons. It has shifted from being regarded more as an unusable means of mass destruction to recognising their practical use for battlefield conflicts. With the improvement of China's nuclear capability, the PLA's confidence in its ability to successfully wage nuclear war has been considerably enhanced. Indeed, younger generation of PLA Generals have gradually broken the psychological shackles imposed by the previous strategy of minimum deterrence. In the conduct of "war games", these Generals often formulate scenarios in which they fire nuclear missiles in high-tech limited war. China has paid special attention to the role of tactical nuclear weapons to assist the conventional forces. A senior PLA General in a recent discussion on the strategic functions of nuclear war, assigned a significant role to the tactical nuclear weapons. He opined "when manpower, firepower and military equipment of conventional troops are short, or when they are reduced to inferiority, development or use of tactical or war-zone nuclear weapons can make up and re-adjust conventional forces. In terms of the positive-defensive strategy, medium-and-short-range missiles with nuclear warheads and tactical nuclear weapons that have actual combat capacity can serve as an actual threat to the enemy's heavy massing of forces and preparations for a large-scale offensive." Data published in the *Nuclear Weapon Databook* indicate that China has conducted at least a dozen underground or atmospheric nuclear tests with a yield below 20 kilotons. The PLA is believed to have developed a tactical nuclear capability in its large-calibre multiple rocket systems (273mm and 425mm) as well as in its new guns (203 mm).

China's strategic community is beginning to realise that the No First Use (NFU) doctrine is a liability as it precludes evolution of an effective deterrence. In any case, Chinese strategists did not consider using nuclear weapons in own territory as violating NFU. The concept of own territory covers Taiwan and may well extend to claimed and occupied areas such as Arunachal Pradesh, Aksai Chin, etc. On the other hand, Chinese Defence literature indicates interest in launch on warning of nuclear weapons or launch under early attack, both of which would scrap the pledge. A book compiled by Chinese Academy of Military Sciences on Nuclear Strategy states:

> "On the basis of the general policy of second strike, the nuclear counter-attack of the future will be implemented through a nuclear counter-attack campaign after the enemy's nuclear surprise attack. The first nuclear counter-attack must strive to be an immediate retaliation, that is, react quickly and rapidly implement a nuclear counter-attack after determining the enemy's missiles are coming, but before they have exploded."

Notwithstanding China's policy of NFU and the inherent contradictions therein, China is acquiring a larger quantity of mobile longer range missiles and the related systems which would cover the continent of United States of America. Further, the focus currently appears to be on space based command and control systems, anti-satellite systems and civil defence infrastructure for damage control. Beijing nuclear force is relatively vulnerable and comparatively far behind that of the United States, and at present far less sophisticated than those of Britain and France. But it does not need to be equal, given the objective of nuclear deterrence. However, when compared to India, China is miles ahead in the field. The Chinese nuclear doctrine has high degree of in-built ambiguity, dynamism and flexibility. It is offensive in nature. The doctrine will continue to get fine-tuned as the Chinese acquire higher capabilities and greater confidence.

Internal Security

Internal security has always been integral to China's national security doctrine. At present the main threat to China is from within and not outside. While the Chinese Communist Party still maintains its grip on political power in China, economic and social liberalisation in the recent years has drastically undermined its role and structure, especially at the grassroots level. The Party's reputation has also been eroded by widespread corruption and nepotism. Corruption has become rampant in China. Approximately 15000 cases pertaining to corruption were investigated in 2001, an increase of 40 percent over the last year. Economic growth has also slowed down. Urban unemployment, mass rural migration and the regional disparities between the developed Eastern coastal regions and the under developed hinterland has the potential to fuel age-old resentments across ethnic, linguistic and cultural lines. The numbers of rural illiterates and school dropouts has increased. Rural income growth has slowed to four percent. With 130 million rural unemployed already, social unrest will only get exacerbated. In terms of modernisation and industrialisation, the party has in effect turned its back on the rural masses that brought it to power. Outbreak of ethnic disturbances and terrorist attacks by separatist groups occur periodically in outlying minority areas of Xinjiang and Tibet. The Muslim Uighur separatists in Xinjiang are fighting for an independent state under the banner of the East Turkistan Movement (ETM). Activists of the ETM met in Munich (Germany) in February 2002 and issued a paper on the human rights situation in Xinjiang. The paper stated that the Chinese had stepped up oppressive policies against Uighurs after 11 September 2001. Some 3000 Uighurs had been arrested and 21 awarded the death penalty. There are indications that Xinjiang and Tibetan separatists groups, duly assisted by Pakistan and Afghanistan based fundamentalists, could join hands to challenge the Chinese authority. More than fifty years after PLA marched into Lhasa, the Tibetan independence movement created a worldwide support network, remains much stronger than it was in 1950. The creation of several new nations on China's Central Asian frontier with ethnic population on both sides of the border has also made ethnic separatism a major concern. The Chinese are very sensitive to any internal problem as is amply evident from their handling of Falun Gong and Zhong Gong movements. Falun Gong is the first mass movement after the 1989 pro-democracy students unrest. The movement banned since July 1999, has been a source of worry for the Chinese government. In spite of harsh laws, the party has not been able to crush it. From January 2001 to February 2002, there have been 83 illegal demonstrations by this movement. Millions of practitioners who combine traditional Chinese meditation with a smothering of Buddhism and Taoism have gone underground lest they be arrested. There were nearly 5000 protests throughout China due to various reasons during the year 2001. The Chinese authorities are very concerned about the various internal developments and have initiated numerous measures to revamp internal security. In 1983, a new Ministry of State Security was created. This was followed up with the transfer of 500,000 PLA troops to the new People's Armed Police (PAP). This process is continuing with PLA formations being converted into PAP at irregular intervals; at least 14 PLA divisions have been transferred to the PAP. Traditionally the Chinese Armed Forces have always been in the forefront in dealing with political wartime contingencies including political turmoil, ethnic unrest and religious or ideological based fundamentalism. However the use of PLA to quell the Tiananmen demonstrations of 1989 became highly controversial within the PLA and there would be substantial resistance to the employment of the Army for such risky and politically divisive contingencies in future. This burden is thus passed on to the PAP. Nonetheless the internal situation in China in the next decade appears to be as mementous, as the last decade was for Russia.

Emerging Scene

At the operational level, the restructuring of the ground forces and the development of improved doctrine propel the PLA's progress towards combined-arms operations. The PLA still has a long way to go toward effective joint service operations, but the new combat regulations firmly establish joint operations as a goal for continued doctrinal development.

More broadly, the reforms represent a major step forward in the regularization of PLA operations. The "first-generation" combat regulations of the 1980s were far too general in their scope to constitute a working operational doctrine. If some of the 1999 Ordinances prove to be as detailed and comprehensive as initially reported, they, together with the comprehensive restructuring of the force and the 10 year logistics reform programme, will represent a major milestone in the professionalisation of the PLA and its slow maturation as a military institution. Despite China's economic success, China's doctrinal evolution suggests that it is emerging as the next status quo global power in the making. China's official enunciation in this regard remains circumspect.

According to China's White Paper on National Defence released in October 2000, the basic objectives of China's defence policy involve "consolidating national defence, resisting aggression, curbing armed subversion, and defending the state's sovereignty, unity, territorial integrity and security". Greater stress seems to be laid on defending sovereignty and curbing internal disturbances. Considering the continuing territorial disputes that China shares with its neighbours, stress on national unification is bound to delay and complicate China's rise as the next global power, be it status quoist or otherwise. China also remains entangled in internal problems that might influence its future course of events, thus influencing its thinking and doctrines. Uncertainties about its influence make China's leaders circumspect in clearly enunciating on China's national security doctrine, though signals of change can be perceived.

The Implications

China's doctrinal evolution may be 'pushed by technology rather than pulled by strategy'. Thus there could be an unbalanced development among three factors of military doctrine, force structures, weapons and equipment. China's evolving military doctrine has the following strategic implications:

(a) 'Active Defence' suggests that Chinese Armed Forces may be prepared to be employed outside Chinese territory in punitive wars.

(b) As China's military capabilities increase, its foreign policies may become more assertive and 'Active Defence' could be invoked for preventing China's containment, foreign intervention or hegemonism.

(c) China's offshore maritime ambitions necessitate the Navy becoming an important instrument of foreign policy.

(d) China views future conflicts as potentially involving short duration, high intensity combat. The PLA's operational focus has shifted from defensive to offensive operations and the need to gain the initiative early in any engagement. Naval and air power is of particular importance in such operations. Thus, even within a military strategy designed to be defensive in nature, force projection is a key result area for China's defence planners.

(e) Till China's nuclear and missile forces become adequately responsive to threats from the USA and Russia, China may continue to exploit existing arms control regimes to extract maximum advantages for itself, may it be political, strategic or technical.

(f) The renewed emphasis on 'active defence', its efforts to enhance the power projection capabilities and the importance it attaches to its land and maritime boundaries clearly indicate that India and Vietnam prominently figure in Chinese strategic calculus. Though there is no threat of direct confrontation with China in the short term, the possibility of a border conflict in long term cannot be totally ruled out.

Chapter 5

PLA's Modernisation

Since the early 1990s, the focus of China's military strategy has been on preparing for military contingencies in the Taiwan Strait and South China Sea. China's military strategy emphasises acquiring capabilities to counter improvements in Taiwan and other regional military forces. It has also been preparing for the capabilities the USA and Japan might bring to bear against China, in the event of a conflict in the Asia Pacific region.

A fundamental objective of China's military modernisation programme is to create a force level sufficient to counter any regional threat. China also seeks to maintain the credibility of territorial claims, protect national interests, maintain internal security and deter any moves by Taiwan towards independence. To support and sustain these forces during times of war, China is also trying to establish a more effective national mobilisation and logistics support system.

If Beijing perceived that war was inevitable, China would attempt to contain and limit the conflict, but fight with sufficient forces and tactics to achieve a military solution before outside powers could militarily intervene and much before vital trade and foreign investment were disrupted. If a third party were to intervene militarily in a regional conflict involving China, the PLA would employ all means necessary in the hope of inflicting high casualties and weakening the intervening party's resolve.

Defence Modernisation Parameters

Priorities : Defence modernisation has been accorded last priority in the 'Four Modernisations', after agriculture, industry and science and technology. The modernisation priorities within China's Armed Forces are focused on Second Artillery Corps (strategic missile force), Navy, Air Force and Army. PLA strategy calls for building these elite forces, which will be able to respond quickly to a regional hot spot and dominate land, air, sea, space, and electromagnetic spheres of battle space. Since 1989, Chinese diplomacy has focussed on institution and enhancement of Confidence Building Measures (CBMs) with India, Russia, Vietnam and Central Asian Republics (CARs). In the short term, there is no compelling land threat to China. Hence, American analysts assess that China has placed a low priority on improving the quality and quantity of its Armed Forces along land frontiers. The resources so released are diverted towards development of maritime, air and space forces. China will continue attempts to seek self-sufficiency in defence technology and production capabilities. Purchases of weapons systems will be made to satisfy critical near-term needs and to gain access to frontier technologies. Nevertheless, the PLA will be heavily dependent on foreign sources for key modern weapons and hardware, platforms, technologies, as well as for systems engineering and integration. Even if the PLA were to acquire the modern weaponry, integrating those systems and training commanders and troops to absorb them will remain a difficult task and inhibit the PLA's maturation into a world-class military force, at least for the next two decades.

Stages : It was against this backdrop that the Central Military Commission had projected the defence modernisation programme with emphasis on professionalism, streamlining and combat effectiveness for the Armed Forces in the following three stages: -

(a) Initial Modernisation from 1992 to 1996.

(b) Fundamental Modernisation by 1998.

(c) Basic Advanced Modernisation by 2001.

Objectives : The objectives of defence modernisation are:

(a) Short Term

(i) Induction and development of new generation of weapons to enhance the combat effectiveness of its conventional forces.

(ii) Enhancement of joint operations capability between the ground forces, the Navy, the Air Force and the Second Artillery.

(iii) Infusion of high technology to improve surveillance and intelligence capabilities.

(iv) Reduction of the PLA strength as a part of downsizing so as to convert it into a smaller but more effective force – a shift from manpower intensive to hi-tech lean and mean force.

(v) To build an elite rapid deployment force within the PLA, equipped and trained in realistic force-on-force scenario.

(vi) To improve the guidance and accuracy of medium and long-range missiles.

(vii) Revamp R&D and modernise defence industry.

(viii) Modernisation of logistics support system and maintenance of adequate strategic stockpile.

(b) Long Term

(i) Reorganisation and modernisation of higher level of Command, Control and Communication system for developing capability to use armed forces as an instrument of military diplomacy and statecraft.

(ii) Enhance strategic power projection capability commensurate with a great power status.

(iii) Having a Blue Water Navy to counter potential security threats in maritime areas from highly sophisticated armed forces.

(iv) Developing space and information technology as a major global power in the 21st Century.

In the opinion of some China experts, important factors in the process of defence modernisation are:

(a) It is an incremental over-all development package to be executed selectively.

(b) Reduction in manpower, with a shift from quantity to quality.

(c) Emphasis on hi-tech weapons and equipment, indigenous as well as imported – and on Research and Development to bridge two generations gap in the conventional and missiles weaponry.

(d) China has one million scientific researchers, of whom one third are involved in defence research and development. Nearly 50 per cent of China's expenditure on scientific R&D is in defence field.

(e) In June 1999, a budget of $ 10 billion was allotted for R&D in seven key areas of space, lasers, automation, information systems, new materials, energy and bio-technology. Of these, the first six have defence applications.

(f) China has exploited Russian defence industries' requirement of dollars to stay afloat. During 1996-2001, China and Russia have signed over 200 agreements and more than 5000 Russian scientists and technicians are working in China's defence industries.

(g) In October 1999, China signed an agreement for military technical cooperation with Russia involving procurement of weapons and technology worth $ 25 billion during 2000-2005.

In October 1995, a national conference was held in Beijing to discuss the implications of the global ongoing revolution in military affairs (RMA). It was then decided that China would specifically target the following :

- China must close an "information gap".
- Network all forces.
- Attack the enemy $C^3$1 to paralyse it.
- Pre-empt enemy attacks.
- Use direct energy weapons.
- Use computer viruses.
- Use submarine-launched munitions.
- Use anti-satellite weapons.
- Forces to prevent a logistics build-up.
- Use special operations raids.

MODERNISATION OF PLA (ARMY)

Strength and Deployment

The total Army strength is around 1.7 to 1.8 million, of which the Infantry troops are about 9,00,000, Armoured Corps around 1,50,000, Artillery about 1,20,000, Air Defense force is approximately 1,00,000, the Garrison troops 3,60,000, professional support elements are about 2,40,000. The Army is equipped with : tanks and infantry combat vehicles - over 14,000, artillery pieces - 13,000, multiple-barrel rocket launchers - over 2,100, and self-propelled artillery - 1,500.

The key points of the PLA's force disposition have been re-diverted to the Southeast Area of Chinese mainland since the signing of the Sino-Russian Bilateral Force Reduction Agreement in 1998, as follows:-

(a) *Southeast Area:* Nanjing Military Region (MR) and Guangzhou MR. Five Group Armies (GAs - each equivalent to a Corps) are deployed, numbering about 4,80,000.

(b) *Northeast Area :* Shenyang MR. The total strength is approximately 310,000, organised into five GAs.

(c) *Southwest Area :* Chengdu MR. The total manpower amounts to 1,80,000, composed of two GAs. Some of these troops are responsible for operations in Tibet.

(d) *Northwest Area :* Lanzhou MR. Two GAs are deployed in the area, a force numbering more than 2,20,000.

(e) *Capital Area :* Beijing MR. The total strength amounts to about 4,10,000 troops, organised into six GAs.

(f) *Strategic Reserve :* Jinan MR. The total manpower is some 2, 40,000 troops, consisting of four GAs.

Deployment in Hong Kong : After merger of Hong Kong, Guangzhou MR was made responsible for positioning troops there. This deployment comprises approximately 6,000 troops with some equipment like armoured vehicles, helicopters and combat crafts. In addition, the PLA has selected Stonecutters Island in Hong Kong as a gate for the construction of a large deep-water harbour to anchor the ships based in Hong Kong. Besides, it may be used as a forward base of the South Sea Fleet as well as a relay harbour in support of the operations at the South China Sea.

Though China's current military buildup indicates an orientation towards Taiwan, the Chinese have adequate forces (Chengdu and Lanzhou Military Regions have 13,14, 21 and 47 Group Armies with additional five divisions) for contingencies vis a vis India and Vietnam.

MODERNISATION PROGRAMMES

Thrust Areas : Army modernisation has a low priority. Force reduction efforts and selective fielding of new equipment focus on building a more mobile, combat ready core within the larger Army. Currently, Beijing is downsizing its Armed Forces. The manpower reduction will contribute to a restructuring of the PLA into a military force consisting of three components:

(a) A small number of hi-tech forces for flexible use in regional contingencies.

(b) A large number of forces equipped with low-to-medium technology weapons for internal security.

(c) A nuclear force to maintain a viable deterrent against other nuclear powers.

Rapid Reaction Force (RRF) : The PLA's rapid-reaction strategy is based on the premise that China will only be engaged in local wars in the near term, and that the PLA must strike quickly to end the war and meet Beijing's political objectives. Furthermore, cost is a big factor, since equipment is becoming more expensive and the cost of maintaining older weapon systems is on the increase. Serious discussions about establishing aviation RRF units did not occur until around the early 1990s, which coincided with China's purchase of the first Russian SU-27s and IL-76s. According to interviews with foreign air force officers in Asia, the PLA AF's rapid-reaction aviation units are currently composed of J-7 (MIG-21), J-8 (MIG-23), and SU-27 fighters and IL-76 transport aircraft. While the bulk of PLA AF consists of J-6 (MIG-19) fighters, the RRF does not include any of these.

Fist Units : During the 1990s, the Army and Air Force have modernised and trained selected formations on priority. These formations are termed as 'Fist Units' or 'Rapid Reaction Formations/Units' (RRFs/RRUs). Such formations/units have regularly practised rapid deployment away from their permanent locations. The rationale of RRF/RRU stems from the emphasis on 'Quality Building'. These units are tasked during exercises on the following aspects:

(a) Surprise attacks on enemy command and control centres.

(b) Disrupt enemy supply lines and distribution system.

(c) Interrupt enemy lines of communication.

(d) Perturb enemy assembly areas.

(e) Conduct armed reconnaissance behind enemy lines.

(f) Cooperate with tactical and operational air assault in the course of Group Armies' main engagement.

(g) Infiltrate and disturb the stability of the enemy rear zone.

(h) Deploy forward observation teams to guide long-range artillery strikes and air support.

(i) Organise and coordinate guerilla activities with the Militia.

15 Airborne Army : According to an October 1993 *Jane's Defence Weekly* report, China commenced converting the 15 Airborne Army's three brigades into divisions, to augment their rapid-response capability. The Kaifeng based 43 Brigade was the first one to be upgraded to a division. 44 Brigade at Yingshan and 45 Brigade at Huangpi were also upgraded to divisions later. Chinese planners had apparently decided that brigade-size forces were too small for combat missions. Some considered opinions about the combat capabilities of 15 Airborne Army are :

(a) The 1999 US Department of Defence assessment describes the 15 Airborne Army as comprising three airborne divisions, each with about 10,000 troops. It is China's primary quick reaction force and has been designated as the National Strategic Reserve. However, its units remain handicapped by insufficient airlift.

Acquisition of additional aircraft and modern equipment, together with the increased emphasis on utilising airborne forces during training exercises, would marginally improve 15 Airborne Army's combat capabilities.

(b) According to a PLA AF Commander, since the PLA AF began receiving several Russian IL-76 transports in 1992, the airborne troops now have all-terrain, all-weather, omni-directional combat capabilities. In order to adapt to various adverse operational conditions, the airborne units have conducted exercises in the snowfields of the Greater Khingan Mountains (Da Xinganling), the hot jungles on the Shiwan Mountains in Guangxi and the Kunlun Plateau, located 4,600 meters above sea level.

(c) In recent years, airborne troops training appear to have focused primarily in and around Tibet. At the same time, however, some airborne training has also concentrated on a Taiwan scenario. By reporting this type of activity, the government is apparently trying to send a signal to inhabitants of Tibet, Taiwan, and Xinjiang that the airborne forces are preparing for internal contingencies, should the need arise. Reporting of these types of exercises also points out some of the airborne forces' limitations. For example, during the 1996 military exercise opposite Taiwan, the Haitan Island, the exercise was scaled down due to inclement weather.

(d) In various exercises, units of 15 Airborne Army have been tasked to:

(i) Seize and hold important areas or landing zones behind enemy lines.

(ii) Cut the enemy's withdrawal routes and intercept reinforcements.

(iii) Cooperate with main forces in destroying the enemy.

(iv) Secure the landing of follow-on forces.

(v) Destroy enemy missiles, nuclear weapons, command and control posts and support units.

(vi) Assist guerilla units and Militia operating in enemy rear.

Army Formations : Chinese brigades normally have about 3,000-4,000 troops and Ground Force divisions have about 15,000 troops. The airborne units are composed of eight types of troops— scouts, infantry, artillery, signals, engineers, anti-chemical warfare and mechanical transport. The Group Armies (equivalent to corps, total 21) are of three types:

(a) *Type A :* Mostly deployed in the North, North East and North West regions; emphasis on armoured divisions and mobility; total strength of each Type A Group Army is about 73,000.

(b) *Type B* : Mostly deployed in the South, South Eastern and South Western regions; crafted for mountainous and irregular warfare; total strength is about 105,000.

(c) *Type C :* Deployed for garrison duties, mostly in the North Eastern region; total manpower 100,000 to 110,000.

Comparison : Available data on the Indian Army and PLA is tabulated for a broad comparison as follows :

	China	**India**
GDP - Purchasing power parity	-$4.5 trillion*	$ 2.2 trillion*
GDP - real growth rate	-8%	6%
Population	1,273,111,290 (Jul 2001 est)	1,029,991,145 (Jul 2001 est)
Defence Expenditure (as percentage of GDP)	-3%	2.7%
Total Armed Forces	-2,310,000 (active) 500 - 600,000 (reserves)	1,263,000 (active) 535,000 (reserves)
Strategic Missile Forces-	-100,000+ ICBM -20+ IRBM -130 -150 SLBM 1 Xia SSBN with 12CSS - N3 (JL - I) SRBM -300	Details not known

	China	India
Army		
	Personnel - 1,600,000	Personnel - 1,100,000
	Main battle tanks - 7,010	Main battle tanks - 3,898
	Towed Artillery - 14,000	Towed Artillery - 4,175
	Self Propelled Artillery 122mm - 1200	Self Propelled Artillery 105mm 80 Abbot; 130mm - 100 152mm - L2S19
	Multiple Rocket Launchers - 2400	Multiple Rocket Launchers - 122mm L 150; 214mm L 12 Pinacha being deployed.
	Helicopters - 321	Helicopters - 162
Navy		
	Personnel - 250,000	Personnel - 53,000
	Submarines - 69	Submarines - 16
	Destroyers - 21	Destroyers - 8
	Frigates - about 42 FFG	Frigates - 11
	Patrol and Coastal combatants - 368	Patrol and Coastal combatants - 39
	Mine warfare - 39	Mine warfare - 18
	Support + misc - 163	Support and Miscellaneous - 32
Naval Air Force		
	Personnel - 26,000	Personnel - 5,000
	Combat Aircraft - 472 shore based	
	Helicopters - 45 armed	
Air Force		
	Personnel - 420,000	Personnel - 145,000
	Combat Aircraft - 3600	Combat Aircraft - 701

Source : The International Institute for Strategic Studies (IISS), *Military Balance 2002-2003* (UK: International Institute of Strategic Studies, 2002)

Infantry Weapons: The modernisation of infantry weapons has concentrated on improving the battlefield survivability, enhanced firepower, targeting systems, cross wind sensors and third generation night sights. The China North Industries Cooperation (NORINCO) and China Precision Machinery Export & Import Corporation manufacture most of the infantry weapons. The NORINCO has 157 factories spread throughout China. The modernization trends are reflected in general reduction of the older versions of pistols, sub-machine guns, rifles, light support weapons, machine guns , anti tank weapons and mortars. These weapons are mostly based on erstwhile Soviet technology. China has taken delivery of unknown number of BMP-3 Infantry Fighting vehicle. In a recently concluded co-production agreement between China and Russia, 10,000 Schmel personal fuel-air-explosive (FAE) projectors are to be build in China. Schmel, a 'flame thrower', has the same effect as a 122mm artillery round. This will give infantry a far more potent rocket grenade that can be used against fortified emplacements and pill boxes. However, sprinkling of Western (Italian and American) and CAR technical know how is also visible. Interestingly, the 5.56 mm automatic rifle M16 A1 was exported to Afghan Mujahideens. Most of the infantry weapons are being exported to developing countries and numerous feuding groups like LTTE and India's North Eastern insurgents. China intends to become a global player in weapons export by 2010.

Armour : China produces various versions of tanks like T-54, T-59, T-69, T-79, T-80, T-85 and T-90. Since 1997, a military technology cooperation agreement with Russia led to transfer of fire control systems of the BMP-3 infantry combat vehicle and its laser range finder. NORINCO is developing Type-90 II tanks, fitted with reactive armour, which is considered to be the most modern tank under development in China and was displayed at the Liberation Day Parade at Beijing in October 1999. A variant of this tank is also being jointly developed with Pakistan (the Al Khalid project, under a 1990 Sino-Pak agreement). The later tank models, from T-85 onwards, have greater speed, endurance, improved accuracy of main gun and night vision devices. Likewise, the later versions of armoured personnel carriers (Type 90 and WZ 551) are improved versions in terms of armour plating, higher power to weight ratio and longer ranges. China has purchased the Russian BMD-3 airborne infantry fighting vehicle.

Artillery Systems : NORINCO has unveiled a new long-range artillery system "Super Range Rocket" which can fire 406 mm rockets mounted on 8 x 8 chassis up to ranges of 360 km. It has also developed tracked and towed howitzers of calibers 122 mm, 130 mm, 152 mm, 155 mm and 203 mm. The China Precision Machinery Export & Import Corporation has developed multi-barrel rocket launchers like A 100 (300 mm) which can fire 10 different types of rockets at ranges of 50-100 km in 60 seconds. The latest acquisition from Russia include Smerch Multiple-Launch Rocket System (MLRS) and Kitolov laser-guided artillery round. Foreign systems are reportedly being sought to give the Army a much better mobile air-defence. China will be purchasing the Tor-MI(SA-15) SAM systems in large numbers which will enhance its short-range air defence capability. The Tor is said to be useful against targets ranging from helicopters to precision-guided munitions. China is also interested in the more modern short range air defence system. The Army has yet to acquire a large attack helicopter. PLA delegations have visited Italy's Augusta and are also holding discussions with South Africa for attack helicopter purchases.

Capability and Prospects

By and large, Chinese Army's weapons inventory remains about 15-20 years behind that of Western and even some Asian Armies. Although it has gone in for 'Quality Building' with gusto, only the RRF (which comprise around 20 per cent of the total force) have the combat capability comparable to modern Armies. Even these lack the airlift or sealift capability vital for power projection. Moreover, the deployment is coast-and Beijing-centric and the majority of its bayonet strength comprises recruits.

The last major campaign in which the Chinese Army participated was the Korean Campaign of the 1950s. The 1962 India-China border conflict cannot be cited as an objective example of PLA capability. In the 1967 border conflict with Russia (then the USSR) and the 1979 'Teaching Lesson' conflict with Vietnam, the PLA did not give a worthwhile account of itself. Ipso facto, the PLA's bread and butter during the last over two to three decades has been suppression of dissent in Tibet, Xinjiang or the student unrest in Tiananmen (June 1989). Hence, it is not a battle tested Army in modern times. Since mid-1990s, the PLA has laid considerable stress on 'Quality Building' and has concentrated on a '21st Century-Oriented Composite System of High-Tech Military Training'. The PLA (Army) is now engaged in combined arms exercises, use of simulators and painting exercise scenarios wherein "Red Army" (PLA) counters "Blue Army" (exercise enemy). In September 1999, a permanent combined force exercise enemy was reported in Shenyang Military Region (North East China). While China's Ground Forces may progressively induct better quality men, weapons and equipment, their absorption by the rank and file is likely to remain a matter of debate. The pace of this aspect will determine the actual impact of defence modernisation on the Ground Forces. Keeping in view the Sino-Russian bonhomie after their Friendship Treaty of July 2001, availability of spare parts of hi-tech weapon systems and technology transfers may no more be a problem for China.

MODERNISATION OF THE PLA NAVY (PLAN)

Strength and Deployment

Current strength of PLAN is estimated to be 2,60,000, according to IISS *Military Balance* 2002-03, of which the Fleet Force is over 1,30,000, Naval Aviation about 26,000, Marine Corps around 10,000, and 40,000 conscripts.

China believes that the greatest potential for sea conflicts might erupt at the East China Sea and the South China Sea. Thus, the "Emergency Mobile Force" is assigned to the East Sea Fleet and the South Sea Fleet. Although the North Sea Fleet has a better equipment, yet generally speaking, the total combat capabilities of the East Sea Fleet and the South Sea Fleet are superior to others :

(a) *North Sea Fleet :* This is deployed in the area from Yalu Jiang, Liaoning to Lianyungang, Jiangsu. This fleet has major bases in Lushun and Qingdao, with over 950 ships.

(b) *East Sea Fleet :* This is deployed in an area from Lianyungang, Jiangsu to Zhaoan, Fujian. This fleet has the major bases in Shanghai, Zhoushan and Fuzhou, with 980 plus ships.

(c) *South Sea Fleet :* This is deployed in an area from Zhaoan, Fujian to Beilunhekou, Guangxi and the Paracel and Spratly waters. It has the major bases in Guangzhou, Zhanjiang and Yulin, with 800 odd ships.

Importance of Naval Modernisation

The Chinese Navy's mission is to protect 18,000 km coastline and an exclusive economic zone of over 3 million sq km, roughly one-third of China's size. Besides a long coastline, PRC claims sovereignty over 7100 islands, out of which more than 30 are yet "in the hands of foreigners".

China claims 80 per cent of South China Sea. Therefore Naval modernisation has been accorded top priority among the three Services. With an economy, which is expected to surpass that of the USA around the years 2020-30, more of China's merchant marine ships can be expected to frequent Pacific and Indian Oceans. Interestingly, while the Army strength has been progressively reduced, Naval manpower increased from 8.1 per cent in 1980 to almost 10 per cent by end 1990s. The Naval share of the defence budget has also increased considerably. The following aspects are noteworthy:-

(a) Chinese indigenous ship building capability is considerable. Shanghai alone has more warship building yards than those in the entire India. The basic Naval technology in China is far more developed than that of India.

(b) Most of Chinese ships are of old vintage and in areas like sensors, weapons, design sophistication and training the Indian Navy has a marginal edge.

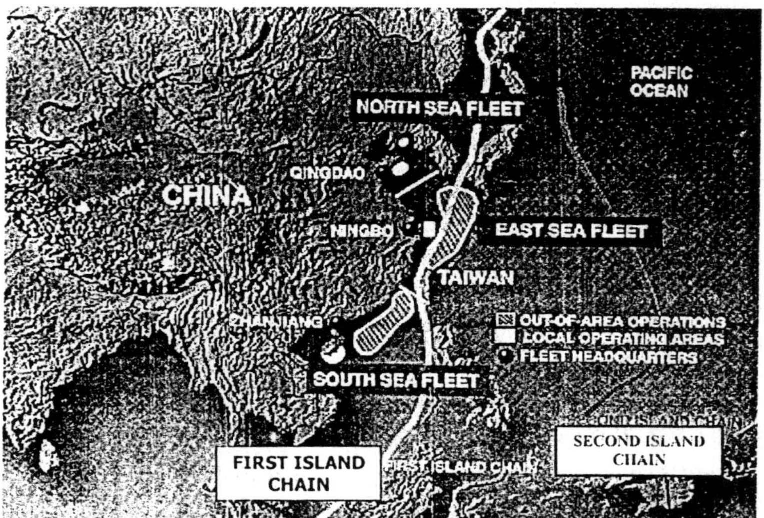

China seeks to transform the PLA Navy (PLAN) from a coastal "Brown Water" (100 nautical miles) to a "Blue Water" force that is capable of reaching the 'Second Island Chain'. This chain is defined as a line from Japan through the Bonin, Marianas and Caroline Islands. The PLAN faces major limitations in its efforts to develop into a power projection force. The Map gives a clearer picture of the 'First and Second Island Chains'.

PLAN Modernisation Plan : In mid 1980's General Liu Huaqing, one of the 'Long March' veterans, formulated a three-phased plan to develop China's Navy. The aim of this modernisation plan was to develop the Navy into a world-class sea power by 2050. Available details of this plan are:

(a) *Till 2000 :* The First Phase focussed on training and enhancing existing formations, renovation and improvement of the conventional naval vessels. (Objective to deter regional threats and to fight battles quickly and at low risk.)

(b) *2001-20 :* The Second Phase will concentrate on the construction of several light aircraft carriers of 20,000 to 30,000 tonnes, purchasing several warships to supplement the carrier task force in order to improve the strength of the fleet and to holster the PLAN's offshore combat capability, projection of the PLAN not only in the Western Pacific but exploring oceans around the world with aircraft carriers and hi-tech equipment.

(c) *2021-50 :* The Third Phase will transform Chinese Navy as a major sea power with 'blue water' capability.

Of special interest to India would be the following Chinese Naval modernisation programmes:

(a) The Chinese interest in the acquisition of aircraft carrier has fluctuated from Russian 'Varyag' through a Macau company, to French 32,000 ton carrier 'Celmenceau' over a period of time. There are unconfirmed reports that China has purchased an 'aircraft carrying cruiser', a term used to describe 40,000 ton Russian Kiev class helicopter carrier. Some Western Sinologists reported that the work on a light aircraft carrier of approximately 30,000 tons is believed to have started. China is planning to have one aircraft carrier by 2010, two by 2015 and four by 2025 A.D.

(b) Six Song-class submarines are to be ready by 2002.

(c) A submarine launched version of C-802 anti-ship missiles is under development.

(d) Moskit supersonic anti-ship missiles and the KA-27 anti-submarine helicopter are being procured from Russia.

(e) In addition to its Marine brigade, PLA Navy is capable of transporting equivalent to one division for amphibious landing. Three Army divisions have also been assigned amphibious operational role.

The Chinese Navy is developing Luhu class destroyers (displacement 4,200 tons) and Jiangwei class guided missile frigates (2,250 tons). Their systems are dependant on Western defence suppliers. By late 1990s, PLA had 60-70 surface combatants including destroyers, frigates, variety of mine sweepers and light patrol craft. China is also acquiring Sovremenny class Russian destroyers, which represent significant upgrade in Chinese Naval capability.

Submarine Warfare : China's subsurface warfare capabilities are modest compared with international standards, but they are considered effective against most other East Asian navies. The Navy's equipment is less sophisticated, older, and noisier. Its personnel are undereducated, the senior enlisted concept is new, and training and exercises lack realism. Currently, the PLAN has access to a wide variety of technology sources and actively engages in technology transfer to further its antisubmarine warfare (ASW) programmes. As China combines domestic research and development with submarine related technology acquired through direct purchase and transfer from foreign countries, particularly Russia, the Naval ASW capabilities are expected to improve. The acquisition of four-KILO attack submarines from Russia has provided the PLAN with access to technology in sonar development, as well as modern weapons systems. China can be expected to incorporate some aspects of these technologies into its domestic submarine construction programmes, although it will take the Navy many years before it can use effectively the advanced technology now available. China's most modern indigenously built diesel attack SONG submarine also incorporates technologies from Russia, as well as from Western countries. The PLAN currently has five HAN nuclear attack submarines (SSNs) in its inventory and one nuclear powered ballistic missile submarine (SSBN) the XIA. Beijing's next generation nuclear submarine programmes are expected to reflect a significant Russian influence. By the middle of the next decade, China is expected to launch its next class of nuclear powered attack sub-marines, the Type 093. China is believed to have good access to a wide variety of foreign sonars, to include passive ranging sonars, flank array sonars, variable depth sonars, as well as helicopters equipped with dipping sonars.

The Navy's mine stockpiles include vintage Russian moored contact and bottom influence mines, as well as an assortment of domestically built mines. China currently produces the EM-11 bottom influence mine, the EM-31 moored influence mine, the EM-52 rocket-propelled rising mine, and the EM-53 ship-laid bottom influence mine which is remotely controlled by a shore station. China is believed to have the acoustically activated remote control technology for its EM-53. This technology probably could be used with other Chinese ship-laid mines including the EM-52. Application of this technology could allow entire mines to be laid in advance of hostilities in a dormant position and activated or deactivated when required. China reportedly has completed development of a mobile mine and may be producing improved variants of Russian bottom mines and moored-influence mines. Over the next decade, China may attempt to acquire advanced propelled-warhead mines, as well as submarine-launched mobile bottom mines.

Anti Ship Cruise Missiles (ASCM) : There are currently four types of ASCM in the Chinese Navy. These ASCMs include the CSS-N-1/SCRUBBRUSH missile carried aboard the JIANGHU I-class frigate (FF), the CSS-N-2/SAFFLOWER onboard the LUDA-class destroyer (DD), the CSS-N-4/C801/SARDINE carried by the JIANGHU III-class FF, and the CH/SS-N-6/C802/SACCADE carried on board various destroyers, frigates, and guided missile patrol boats (PPGs). The C802 is an extended range version of the solid fuel C801 missile. The SS-N-22/SUNBURN supersonic ASCM system will be deployed on the two SOVREMENNY-class guided missile destroyers purchased from Russia.

Capability and Prospects

The Chinese Navy remains a coastal defence force. The American analyst David Shambaugh has opined that by late 1990s, it had graduated from 'brown-to-green water capability with minimal blue water capacity'. Other analysts feel that it will take considerable time and expenditure to build a blue water Navy of significance. Chinese Navy's most serious shortfall is a lack of air power at sea. When compared to Western Navies, it also lacks in all categories affecting modern naval operations training and education, naval systems and platforms, naval warfare technology and systems, doctrine and tactics, command and control, intelligence, strategic planning, operational expertise, and influence within the national decision-making structure. Despite recent purchases from Russia and overall improvements, there is no evidence to support that the Navy has initiated major modernisation effort required to transform the coastal navy into a power projection force by 2020.

China is likely to remain a net oil importer. It has been enforcing claims to prospective oil rich South China Sea. This, coupled with the increasing demand for cheap Gulf oil, will ensure that China's quest for a 'Blue Water' Navy by around 2020 may get accelerated with passage of time. Whilst the present Chinese Navy has a considerable edge over Indian Navy in size, the latter has much more experience in operating aircraft carriers and relatively sophisticated surface ships and submarines. This edge is likely to remain with Indian Navy for at least a decade or so.

MODERNISATION OF THE PLA AIR FORCE (PLAAF)

Strength and Deployment

The Air Force strength of over 4,50,000 is organised into Aviation troops - 1,40,000 (excluding Naval Aviation), Air Defence - over 1,00,000, Airborne troops - approximately 30,000, and other professional support force - about 70,000. The Air Force, excluding the Naval Aviation, comprises around 3,750 combat aircraft of all types and varying vintage.

The concept of the deployment is to set up one air command at each military region. At present, stationed at the air bases within 500 nautical miles from Taiwan are 1,300 aircrafts, of which some 600 airplanes has a radius of operation over Taiwan proper. In addition, the SU-27 with a radius of operation of about 810 nautical miles, now based at Wu Hu, Anhuri Province and Juixi, Guangdong also covers Taiwan.

Status : As opined by a Taiwan AF officer, the PLA AF is undergoing a "crucial transition from an obsolete giant with short operational legs and limited all-weather capability to an offensive-oriented force with extended range and greater lethality". As per an American Department of Defence report, the PLA AF has over 2200 J-6 (MIG-19 equivalent) obsolete fighters of pre-1960s vintage, several hundred J-7/J-8 fighters (MIG-21 equivalent/upgraded versions) and lesser number of SU-27s. Virtually all domestically manufactured combat aircraft are based on technologies of 1960s.

Modernisation Trends

China's acquisition of Russian aircraft and Israeli combat aircraft technology are pushing China's air force closer to the goal of an all-weather power-projection air service. PLA's Air Force modernisation is focussing on six combat capabilities that is air superiority, ground attack, air transportation, airborne early warning and recce, Electronic Counter Measures, maintenance and logistics. 74 SU-27 Russian aircrafts have already been procured and a decision for the license production of 200 such aircrafts over a 15-year period with a $ 2.5 bn initial contract is indicative of their resolve to modernise their fighter aircrafts. It is also acquiring advanced SU-30s. China is planning to produce FC-1 light weight Jet fighters in collaboration with Pakistan to replace its F-6 and F-7. Egypt is also being co-opted in this project to make it a Chinese-Pak-Egyptian joint venture. The FC-1 is about the size of an F-16, though much cheaper. China may obtain about 100 of these aircrafts. This aircraft may enter PLA AF by 2005-2006. China has also developed the J-10 advanced fighter with Israeli technical assistance and the first test flight would be made within this year. China is also negotiating a deal with the USA for purchase of C-130 Hercules aircraft. There are indications that China may acquire up to four

Russian Antonov An-124 heavy lift aircraft that carries more than US C-5 transport. By 2010 AD, PLA Air Force would perfect air-to-air refuelling capability. Besides converting up to 20 H-6s into aerial tankers, it may go for additional IL-78 refueling tankers. It is reported that China had purchased four IL-78 for $ 1 billion. One refueling extends the radius of the SU-30 from 932 to 1615 miles. At a 600 mile radius, one IL-78 can carry enough fuel to refuel at least six SU-27 or SU-30 aircraft.

Despite improvements across the board, the PLA AF still lags far behind its modernisation goals. It retains a mixed fleet of modern aircraft, such as the J-11 (SU-27), alongside 1950s vintage aircraft. Its ability to conduct all-weather flying and extended operations essential to support an offensive strategy are questionable. Further, China's ability to exploit foreign advanced technologies and equipment in production and utilisation of advanced aircraft is a limitation. However, PLA AF shortfalls will not inhibit its dedication to completing its assigned missions, nor do its shortfalls reassure regional powers of China's long-term regional capabilities and intentions. The Asian countries watch PLA AF developments with growing concern that it could develop sustained combat capabilities within 15-20 years. However, lack of operational aerial refuelling capability and airborne warning and control systems (AWACS) will limit the range of China's air power projection capability.

Prospects

By 2005, PLAAF will have around 2200 fighters, 500 ground attack aircraft and 400 bombers. Majority of them will be second and third generation aircraft, supplemented by a limited number of fourth generation fighters/bombers. By 2010, PLA AF will be a 'leaner and meaner' force comprising 1500-2000 aircraft. This will be mostly J-7s (MIG-21), J-8s complemented by SU-27s and a lesser number of SU-30s. The J-10 (based on Israeli Lavi/US F-16 derivative technology) will most probably be PLA AF's frontline fighter during the next decade. It is also reported that Russia and Israel are competing to offer radar, countermeasures and weapon packages to equip the J-10. China eventually may build up to 300 J-10. It is also appreciated that China may acquire improved C^4I and mid air refuelling capability, which could increase its regional combat potential. By 2020, the PLA AF will have about 400 SU-27 and SU-30 hi-tech fighter aircraft, which could make it the mightiest air power in South East Asia. China may also be seeking to purchase the Russian Tupolev TU-22 Backfire intermediate range bomber for replacement of its aging H-6 (TU-16 Badger) bombers.

The Indian Air Force, working on a truncated requirement of 35 squadrons against an original force level of 45 squadrons, would require to replace over 300 modern aircrafts to maintain a reasonable parity during the next decade, if it is to maintain the existing operational edge over PLA AF. The availability of funds for such a massive acquisition/ production is in doubt.

Acquisition Plan

In order to keep abreast with the latest weapon systems, China has purchased many state-of-the art weapons or sub systems from different countries. China has also purchased system components to improve existing weapons or to help build new military hardware. The most important consequence of its growing access to foreign military technology is the acceleration of PLA modernisation programmes. Though the major source of acquisitions is Russia, it is diversifying its procurement sources. Its latest arms deals with Israel and France are of interest. The details of weapons which China has acquired or has shown interest in acquiring are given on the next page.

Known and Projected Chinese Foreign Weapons Acquisition

Source	Purchased or Used	To Purchase	Acquired or Acquiring	Shown Interest in Acquiring
Russia	MAZ IRBM TEL (1)	SU-27SK co-prod (200-300)	SS-18 technology	ABM technology
	S-300 PMU-1(160+)	Phalcon AWACS (6-8)	SU-30MK (50)	ASAT technology
	SU-27SK (50)	R-77MM	BMD-3 or BMD-IP IFV	Military laser technology
	IL-76	Kh-31 ASM	Smerch MLRS	Kh-65 SE med-range cruise missile
	R-73 AAM	Sovremenny Destroyer (2)	Bazalt Antiarmor Submunitions	Yakhont anti-ship.cruise missile
	R-27 AAM	Moskit ASM	Nona-SVK mortar	RADARSAT technology
	Kh-31 ASM (5)	SA-N-7/12 SAM	Gran laser-guided shell	PGMs
	Kilo SSK (4)	KA-27 ASW helo (4-6)	Kiev-class carrier	300V SAM/ATBM IL-78 aerial tanker
	Mi-8/17 (50+)	APR 3-E ASW torpedo	Kitolov-2 laser guided shell	
	SA-15 SAM	Shmel FAE co-prod	T -90U main battle tank	
	EW/SIGINT equipment		BMP-3 infantry fighting vehicle	
			RADAG technology	
			SSN technology	
Israel	Lavi Techology	Phalcon AWACS*	Delilah cruise missile tech	Military laser technology
	Python 3 AAM			Python 4 AAM, helmet display
France	Crotale SAM			RADARSAT technology
	Dauphin (50)			Aircraft carrier
	Gazelle (8)			
	Ecureuil (10)			
Britain		Searchwater AWACS* (6-8)		GEC-Marconi Argus AWACS
United States	Ecureuil (10)			
	Blackhawk (24)			

AAM - air to air missile; **ABM** -anti-ballistic missile; **AEW** - airborne early warning radar; **ASAT**- anti-satellite; **ASM** -anti-ship missile; **ASW** -anti-submarine warfare; **ATBM** -anti-tactical ballistic missile; **AWACS** -airborne warning and control; **FAE** -fuel air explosive; **IRBM** -intermediate range ballistic missile; **MLRS** -multiple launch rocket system; **RADAG** -radar digital area guidance; **PGM** -precision guided munition; **SAM** -surface to air missile; **SSK** -conventionally powered attack submarine: **SSN** -nuclear powered attack submarine; **TEL** -transporter erector launcher; **SIGINT** -signal intelligence

* *The Israelis are going ahead with this deal despite pressure from the USA.*

Source: James R Lilley and David Shambaugh, *China's Military Faces the Future*, p.87.

STRATEGIC MISSILE FORCE

Strength and Deployment

The Second Artillery Corps, raised in 1965, is the smallest service in PLA with around 4.5 percent of its total manpower. Yet it assumes a disproportionate share of burden of military deterrence through its capability to launch a second strike, nuclear or conventional against the global powers. However, it handles the third largest nuclear missile arsenal in the world. In the Golden Jubilee Parade on 10 October 1999, over 50 percent of the displayed weaponry comprised rockets and missiles. The name Second Artillery was coined by Zhou Enlai, primarily for keeping China's nuclear forces secret to the domestic and international watchers. It has come a long way since then.

This strategic missile force, with a strength of 125,000, is composed of 50,000 Missile troops, Technical-equipped force: some 20,000, Construction troops, Anti-Chemical and Communication troops: 50,000 persons approximately, some 10,000 persons as Training Units and about 20,000 Logistics Support troops. It is equipped with strategic and tactical missiles of various types. Most Western analysts put China's deployed nuclear warheads at approximately 400 including 20 ICBMs, 140 IRBMs and 14 SLBMs (one SSBN).

The strategic disposition of the China's missile forces mainly aims at the United States, with only a small portion targeting the neighbouring states. The disposition of missiles is deployed in a small number and scattered throughout the Chinese Mainland so as to increase its survivability in a nuclear war. At present, China has completed its disposition of DF-3 (Dongfeng 3), DF-4, DF-5, DF-15 and DF-21 of which the DF-15 (M-9) tactical mobile missile is deployed in the area of Leping, Jiangxe, while its forward positions are scattered along the Jiangxi-Fujian area with ranges covering whole of Taiwan. PLA has approximately 300 DF-15 with a range of 600 km having a conventional warhead of 500 kg and nuclear warhead of 20 kt. The PLA's missile units have built several hundred underground tunnels, living accommodation and storage facilities in the most inaccessible areas in China.

The Chinese have vigorously perused strategic modernisation to improve the safety, survivability and reliability of nuclear arsenal. Currently about 14 per cent of China's strategic missiles can reach the USA. PLA feels that if this ratio is raised to 70 per cent, it will effectively strengthen national security against Western powers. In fact long range capabilities are considered as a key factor of deterrence. The aim has been to extend missile range to reach potential targets from the US bases in Japan and Korea to Guam and continental America.

The PLA is upgrading short-and medium-range missile forces for use in conventional combat, as well as developing strategic and tactical missiles. New ballistic and cruise missile systems are being developed. Where possible foreign technology is being sought to improve China's future missile development. Increasingly, Chinese missile forces will use non-nuclear warheads. By the year 2010, according to a 1996 report by the U.S. Department of Defence's Office of Net Assessment, China could have as many as 2000 conventionally armed missiles of varying ranges, some of which could reach as far as Guam. However, a 1997 Department of Defence report to the Congress notes that China has the potential to build "as many as a thousand" new ballistic missiles over the next decade. The two ICBM development programmes and one submarine launched ballistic missile programme are being vigorously followed which may result in new deployed missiles by 2010. The 1125-mile range DF-21 IRBM is also being improved. The solid fuelled missile is both road and rail mobile. Sources in Taipei reported that China is working on 1700-mile range and 1800-mile range variants of the DF-21. It is yet not known when China will be able to complete either a RADAG System, believed to have been given by Israel, or longer range DF-21. The East Wind (Dongfeng) ballistic missile series (especially the DF-15, 21, and 25) are being adapted for use in limited conflicts. The new DF-31 with 5000-mile range (a solid-fueled, mobile, long-range intercontinental ballistic missile [ICBM] was successfully tested in 1995. The DF-5A ICBM is being replaced by DF-41 (range 12,000 km, mobile, three stage solid propellant) ICBM which can reach almost anywhere in the world. China is likely to increase its ICBM force to 30. It intends to base these missiles increasingly with fixed launchers in hardened silos, using solid fuel propellants, and hopefully outfit as many as possible with multiple independently targetable reentry (MIRV) warheads. However, China has not yet mastered MIRV technology, though in early 1998 U.S. Air Force General Eugena Habingar stated that China has made considerable progress in developing MIRVs for its ICBM. China's intermediate-range ballistic missile (IRBM) force is also being modernised, while improvement of submarine launched ballistic missiles (SLBMs) is also accorded high priority. The Second Artillery Corps seeks to develop a new generation of warheads that are smaller, "cleaner" (less collateral damage) and more accurate. China normally conducts missile tests over its land territory, at reduced range. It is hoping to narrow the technological gap with nuclear powers and achieve a circular error of probability of under 50 metres. A recent report from Hong Kong notes that PLA units are using a Differential GPS system which covers the Republic of

China (Taiwan) with a 1 to 3 meter location accuracy. China had bought a Russian MAZ transporter-erector-Launcher (TEL) to upgrade Chinese TELs with advanced suspension and navigation systems. A more mobile TEL would enhance the survivability of DF-31 or DF-41 ICBMs.

It may be recalled that the USA had fired Cruise missiles at Osama Bin Laden's suspected hideouts in Afghanistan during mid 1999, on the basis of intercepts of his satellite phone transmissions. Some of these missiles had not exploded on impact and the Taliban made them available to Chinese missile scientists and technicians for inspection, courtesy Pak Intelligence. To what extent the Chinese will be able to utilise such exposures remains an unanswered question, but the Chinese are unlikely to miss out on opportunities which would help them advance their reverse technology programmes.

Capability and Prospects

American analysts consider China as having a modest deterrent with limited second-strike capability. In their view, Chinese strategic forces have certain drawbacks like poor accuracy and slow and unreliable reaction times. However, such analyses are more relevant to a Sino-US showdown. Chinese nuclear and missile capability is far behind the USA and Russia and less sophisticated than that of UK and France. However, it is way ahead of the emerging Indian nuclear and missile capability. Whilst it is debatable whether or not Chinese have stationed surface-to-surface missile in Tibet, the range of their inter-continental and intermediate range ballistic missiles enables them to cover all major Indian cities. This has been amply highlighted in the Indian Ministry of Defence Report (2000-2001).

The Cox's Report brings out that China has managed to acquire information on the design characteristics of the sophisticated W 88 warhead used by the US Navy's Trident Missile. China is also said to have gained critical information on improving the accuracy, range and payload of its missiles. It can thus be inferred that, in future, China may be able to deploy more reliable, accurate and quicker reaction time missiles. The Second Artillery is not yet armed with strategic cruise missiles. Therefore development of land attack cruise missiles for theater and strategic missions have a high development priority. China has shown interest in combined GPS/Inertial and Terrain-Contour Matching (TERCOM) guidance systems for cruise missiles. Russia and Israel are potential technology sources that could help China. It is believed that China has provided financial support to Israel to build a land attack 240-mile range version of the Delilah cruise missile, with a penetrating warhead to be used against hardened targets. The PLA is likely to continue advocating increase in nuclear and missile inventory, more so now that the Bush administration is proceeding with NMD/TMD

PEOPLE'S ARMED POLICE (PAP) AND THE MILITIA

PAP : The PAP, which is also a part of China's Armed Forces, numbers 1.1 million plus, has military and security missions. The Armed Police has six type of formations/units, namely security guards, border defence troops, fire teams, traffic, public utility and mobile divisions. In peacetime, it is responsible for the internal security in coordination with the public security units, the national-level construction works and guards. At wartime, it is under military command to carry out missions such as battlefield security, anti-infiltration and coast guards.

Organisation and Deployment : Since the Tiananmen Square students unrest, the status and importance of the PAP have been rapidly elevated. The Central Military Commission has made public that in case there is any domestic upheaval, the PAP will be first mobilised. In other words, the Armed Police is China's main force in charge of the internal security and social stability: -

(a) *Command System :* There are two chains of command for People's Armed Police. The Armed Police is simultaneously under the command of the CCP's Central Military Committee and the State Council. There are reports that the command system of the People's Armed Police is going to be transferred from under the Ministry of Interior to the Defence Ministry and its military function is beefed up.

(b) *Future Trends :* China's Armed Police, having a mixed function between the military police and the police of the democratic countries, is equipped with the light weapons. In recent years, its forces have been beefed up with armoured vehicles, command vehicles and patrol cars, etc. Besides, the air patrol units like the flyboat and helicopter units have been activated with the capabilities in the field of the mobile and long-range patrolling. Its contingents have capabilities better than those of the public security police in general. Some of the GHQ-level units of the Armed Police are given the PLA's military training and exercise, functioning as the main force for internal security duties.

Militia and Reserve : People's Militia and Reserve are the main reserve strength of the PLA's mobilization. The Reserve is mainly composed of the retired and demobilised officers and soldiers, numbering more than 1 million persons. The People's Militia consists of the members of state enterprises, production units and organisations at districts, towns and villages, through a "two-in-one (labour and military)" alignment. Training of Militia/Reserve troops is either conducted in-house or along with the PLA. The training duration is shorter and training standards are lower.

INFORMATION AND COMMUNICATION SYSTEMS

Command, Control, Communications, Computers and Intelligence (C^4I)

China has an extensive network of hardened, underground shelters and command and control facilities for its military and civilian leadership. Fear of a possible war with the former Soviet Union in the 1960s and 1970s prompted Beijing to expend considerable resources constructing national level command posts, civil defence facilities and associated communications. These facilities are intended to ensure survival of China's leadership and provide a refuge from which it can maintain a control over the country's military forces. Both civil and military communications networks support these facilities. The military communications network is probably separate from the civil telecommunications network. It is possible that the two could be linked, a capability the PLA would be expected to exploit in time of crisis. China's military national level command and control communications are carried over multiple transmission systems in order to create a military communications system which is survivable, secure, mobile and less vulnerable to exploitation, destruction or electronic attack. China's communications networks are capable of supporting PLA military operations within China's borders, while they could be degraded, they could not be completely denied.

C^4I modernisation and automation has been a top Chinese priority since 1979. This effort has produced a command automation data network capable of rapidly passing operational orders down the chain of command and moving information to national and theatre level decision makers. However, China's C^4I infrastructure, including the command automation data network portions, is not capable of controlling or directing military forces in a sophisticated, western style joint operating environment. The command automation data network is capable of supporting PLA peacetime operations within China's borders. It can also support limited preplanned conventional attack options along China's periphery. China still lags far behind western standards for controlling complex joint operations and lacks the robust C^4I architecture required to meet the demands of the modern battlefield. China's continuing deficiencies in this critical area are reflected in the failure of its recent second generation military communications satellite, the DF H-3, which failed to become operational following its May 1997 launch. This was the second malfunction, the first occurring in 1994, when the satellite failed to achieve proper orbit.

In October 1998, Beijing Military Region (MR) used a "military information superhighway" for the first time at joint defence warfare drill. This "military information superhighway" was described as an information network subsystem of the "campaign training (command) automation system" developed by the Beijing MR. The system processed and transmitted graphics, characters, and audio data, modernizing the traditional Chinese methods of drill and command. At the MR's operations center, commands from the MR commander and political commissar and military situation maps were relayed to and from the subordinate Group Armies (GA) operational facilities hundreds of kilometers away. Lanzhou MR military leaders reportedly made use of video teleconferencing during a 1997 command post exercise. The teleconferencing linked participants stationed at distant points throughout the MR. It is also reported that China conducted its first operational level logistics support exercise using a computer network in late 1997. The exercise allegedly involved 150 computers and linked 22 separate divisions for the exercise. The eight-day exercise was conducted on an unprecedented scale, connecting units thousands of kilometers apart.

Microwave communication equipment is being used in installations belonging to all branches of the PLA. Chinese open sources report that the 2nd Artillery has a new digital microwave communications system to support its missile launches. The 2nd Artillery signal unit reportedly started developing the system around 1995. This new system reportedly provides the 2nd Artillery with all weather and encrypted communication ability. Cellular telephone services does not yet appear to be a significant element of PLA military communications, although China has shown interest in establishing dedicated military cellular systems for PLA use. However, with the explosive growth of cellular communications in civilian sector, cellular communications could become a significant element of PLA communications in future. China's military communications network continues to be upgraded, however, the bulk of China's military communications reportedly is processed on communication lines run by the Ministry of Information Industry. Both networks are composed largely of commercial off-the-shelf technology. This technology either is not restricted for sale to China or recently was decontrolled, although Beijing probably has been successful in obtaining some restricted technology to modernise its telecommunication network. Europe, Japan and Israel compete to sell telecommunications technology, as well as related hardware and software to China.

Electronic Warfare (EW)

Though nothing much is known about the exact nature of Chinese EW capabilities, Professor Desmond Ball wrote, "China maintains by far the most extensive signals intelligence capabilities of all the countries in Asia/Pacific region" (Signals Intelligence in China, *Jane's Intelligence Review,* August 1995, p.365). The inventory of Chinese EW equipment includes a combination of 1950-1980s technologies with only a few select military units receiving the most modern equipment. The Soviet Union was an early source for Chinese signal intelligence equipment during the 1950s. The Chinese breakthrough came in the early 1980s from cooperation with the United States to establish signals intelligence stations on the Soviet borders to monitor missile tests. This cooperation presumably lasted for a decade or so. In the recent years, the thrust of China's EW efforts continue to focus on technology development and design improvement accomplished mainly through cooperation with Western companies through reverse engineering efforts besides the direct procurement of foreign systems. China is procuring state-of-the-art technology to improve its intercept, direction finding and jamming capabilities. More recently Israel has sold China Signals and electronic intelligence equipment particularly for use in ELINT aircraft including EL/L-8300 SIGINT system. Despite China's best efforts, the sophisticated Western technology is hard to come. With the changed global geo-strategic scenario, Russia continues to remain for China, the traditional source of both EW and electronic intelligence equipment. China is gradually improving the electronic sub-systems on their SU-27 planes. Many of them procured in 1997 are fitted with ECU pads that automatically detect and jam radar signals. Beijing unmanned aerial vehicle programme is making rapid progress. In addition to providing extended imagery reconnaissance, surveillance and electronic intelligence collection, it will probably yield platforms for improved radio and radar jammers. Further, the existing earth stations can be modified to interface with satellite communications. The PLA is also developing an electronic countermeasures (ECM) doctrine and has recently conducted structured training in an ECM environment. In 1998, a Hong Kong magazine believed to have close ties with the Chinese military claimed that PLA could cause US F117 stealth fighters to malfunction and jam Taiwan's anti-ship missiles. Though the veracity of these claims cannot be verified, there is hardly any doubt that the Chinese capabilities in this field are considerably better than most of the Asian countries.

EXPLOITATION OF SPACE

China has the capability to launch military photoreconnaissance satellites; however, the technology employed is outdated by Western standards. Beijing does not possess a real time photoreconnaissance capability, but eventually may deploy advanced imagery reconnaissance and earth resource systems with military applications. The China-Brazil Earth Resources Satellite (CBERS) was launched in October 1999, and the experience gained from operating this satellite supported Beijing's efforts to develop improved military reconnaissance satellites. CBERS also will provide some useful military data. China may also attempt to deploy a near-real-time electro-optical imaging satellite within the next decade, as well as a high-resolution film-based photoreconnaissance satellite. In the interim, Beijing can be expected to exploit commercial SPORT and LANDSAT imagery. Use of other commercial higher resolution satellite imagery also can be anticipated, as it becomes available. China already has launched three low-orbit meteorological satellites and a geo-synchronous weather satellite. Although

Beijing has received some degree of foreign technological assistance in the areas of reconnaissance, surveillance and targeting capabilities, many of its system development efforts appear to have a substantial indigenous component. In future, however, Beijing could be expected to acquire and incorporate greater amount of foreign technology and hardware to expedite programme development. In its efforts to develop a combined 'Reconnaissance-strike complex', China is covertly and overtly acquiring Russian and Western technology to build more progressive imaging and radar satellites and perhaps electronic intelligence satellites. China is interested in 400-500 kg satellites and plans an oceanographic research satellite Haiyang 1 (Ocean-1) in this class that was scheduled for launch in 2001. Other missions for satellites of this class that Beijing eventually may field include earth observation, communications, and navigation. China also is developing minisatellities (weighing less than 100 kgs) for missions, which include remote sensing and networks of electro-optical and radar satellites. A joint venture between China's Tsinghua University and a British University is building the "Tsinghua" system, a constellation of seven mini satellites with 50-metre resolution remote sensing payloads. Subsequent satellites in the series probably will have improved resolution. In addition, Beijing participates in the Asia Pacific Small Multi Mission Satellite Project as part of the Asia Pacific Multilateral Cooperation in Space Technology and Applications Programme, which reportedly includes Iran, Pakistan, Thailand, Mongolia, South Korea and Bangladesh.

Space Launch Programme

Although China is improving its overall space launch programme, there is no evidence that it currently is developing the capability to conduct "launch-on-demand space launch operations," ie, the capability to use satellites and space launch vehicles in storage to launch within 24 hours of a decision to do so. Exploitation of space to include manned space operation remains a high priority. Although nearly all-major aspects of China's manned space programme began within the last five years or so, Beijing is still aiming for a possible first manned launch by 2002-03. While one of the strongest motivations for this programme appears to be political prestige, China's manned space efforts could contribute to improve military space systems in the 2010-20 timeframe. In addition to scientific and technical experiments, Chinese astronauts, for instance, could investigate the utility of manned reconnaissance from space. China is said to be acquiring a variety of foreign technologies, which could be used to develop an anti-satellite (ASAT) capability. Beijing may have already acquired technical assistance that could be applied to the development of laser radars used to track and image satellites and may be seeking an advanced radar system with the capability to track satellites in low earth orbit. It also may be developing jammers, which could be used against Global Positioning System (GPS) receivers. In addition, China already may possess the capability to damage, under specific conditions, optical sensors on satellites that are very vulnerable to damage by lasers. Beijing also may have acquired high energy laser equipment and technical assistance, which probably could be used in the development of ground based ASAT weapons. Given China's current level of interest in laser technology, Beijing probably could develop a weapon that could destroy satellites in future. Although specific Chinese programmes for laser ASAT have not been identified, recent media reports indicate an interest in developing the capability, and Beijing may be working on appropriate technologies.

China has extensive space related cooperation programmes with many countries. Although most of these projects are described as scientific or civilian in nature, military significant technology transfer nonetheless occurs in many of them. According to press reports, Moscow and Beijing have 11 joint space projects underway. These include cooperative manned space activities. The Chinese also have strong interest in Western, Canadian and German radar satellite capabilities, to include a possible purchase of synthetic aperture radar (SAR) satellite systems. China recently signed a contract to launch an Italian built communications satellite with 28 ku-band transponders; two earlier contracts involve research into "observation and data detection satellites" to be built jointly by both countries. China's Xian Satellite Control Site and the French national space agency announced a programme of cooperation for satellite command and control in February 1999. A December 1998 press report stated that China has signed intergovernmental agreements and memoranda on aerospace cooperation with the United States, Canada, Germany, Italy, France, Britain, Russia, Pakistan, India and Brazil. In addition, Beijing has promoted technical and economic cooperation and exchanges of different types, including jointly developing satellites, with corporations, enterprises, and research institutes in more than 70 countries and areas. Since 1998, China and the United States have signed three intergovernmental agreements on launching services. Several US satellite-manufacturing companies have signed agreements on commercial satellite launching services, involving a total of some 30 satellites. The satellite "Sinosat", jointly developed by China, France, and Germany, was launched successfully in 1999. Morever, in the form of a joint venture, China and Germany have made improvements to the DF 3 communications satellite and have worked on a new generation of similar satellites.

According to December 1998 South Korean press reports, South Korea and China are expected to share data and information collected by their respective remote sensing satellites. In November 1998, the Chinese media reported a jointly funded contract with Holland for a cooperative project to develop and use a new satellite to monitor desertification and crop yields in China. Chinese scientists likewise have been reported studying mini satellite technologies in Great Britain. Chinese and British entities have established a joint venture to build and launch China's first privately built satellite. China's aerospace industry also is seeking to integrate GPS and Russian Global Navigation Satellite System (GLONASS) guidance technology into fighters and helicopters. The China Aerospace Corporation displayed a GPS receiver at an exhibition in Beijing in September 1996, and provided brochures advertising both a 12 channel GPS receiver and a 12 channel GPS/GLONASS receiver. One brochure showed a space launch vehicle suggesting GPS use in missile applications. Information obtained at a more recent air show indicates that all of China's new fighters will incorporate GPS navigation systems. China's military backed industries also have entered into joint ventures with foreign firms to produce GPS receivers, which may find their way to military weapons. To complement GPS/GLONASS navigation aids, China has been attempting to acquire commercial satellite imagery from various foreign countries. This widely available satellite imagery could be used in conjunction with GPS/GLONASS to develop digital terrain maps for targeting, missile guidance and planning.

OTHER MILITARY TECHNOLOGY DEVELOPMENTS

Anti Radiation Missiles

China is developing two anti radiation surface-to-air missiles (SAMs)- the FT-2000A. The former SAM probably utilises the yet-to-be-deployed HQ 9 an SA1-/Patriot class SAM while the latter apparently uses the comparatively older HQ 2/CSA-1 SAM, first seen in the late 1950s.

The missile associated with the FT-2000 is probably similar to the HQ-9 in size but may be somewhat slower. Its maximum range reportedly is about 100 km, with a maximum altitude of approximately 20 km. A complete FT-2000 battalion probably would consist of a command platoon and three batteries. Sales brochure acquired at Farnborough dubbed the FT-2000 an "AWACS killer". The FT-2000 could be deployed as a stand-alone air defence system or deployed as part of another system. Although Beijing has stated publicly that the FT-2000 will be available for foreign export around the year 2002, it is more likely that it will not be available until the second half of the decade. If a foreign customer cannot be found, development could take longer or not occur at all. According to a China sales brochure, the FT-2000A SAM will use a highly-modified HQ-2/CDA-1 missile that has been equipped with a passive radio frequency homing seeker operating in the 2 to 6 GHz band. The FT-2000A also will contain a new millimeter-wave band fuze, a new guidance and control section and a new 60 kg fragmentation warhead. The missile has a cutoff valve for thrust adjustment while in flight, probably to extend its range. The missile seeker is loaded with the target aircraft's radio frequency (RF) signature before launch and relies on this information for tracking and intercept. The missile has an estimated maximum range of 60 km, with a maximum altitude of 18,000 metres.

A stand alone FT-2000A battery consists of a central control station and twelve launchers, each holding one missile. The central control station has one master passive sensoı and three auxiliary passive sensors. The auxiliary passive sensors coordinate with the master passive sensor through triangulation to determine angle and range of targets emitting in the 2 to 6 GHz band. This configuration is totally passive relying on the RF emission of the target.

Anti Missile Defence

At a political level Beijing objects loudly to the Theatre Missile Defence and National Missile Defence programmes of the Unites States, but says much less about its own anti-missile programmes and its anti-satellite and space warfare plans. China's ability to defend against cruise missiles is considered extremely limited, while its anti-theatre ballistic missile capability (ATBM) and its antiballistic missile (ABM) capability is considered non-existent. However, Beijing reportedly is developing state-of-the art SAMs, which over the next 10-20 years will improve its ability to defend against cruse missiles and TBMs; though, no significant advances with respect to developing ABM capabilities are anticipated during this period. Most of Beijing's developmental SAMs appear to be derivatives of existing systems. In 1991, China took a major step toward developing a missile defence capability when it reportedly purchased the Russian Fakel S-300 PMU (SA-10) surface-to-air missile (SAM). These Russian missiles perform about the same as early U.S. *Patriot* surface-to-air missiles. According to some U.S. sources, China also has produced the HQ-9 surface-to-air missile from the Russian and US technologies. The HQ-9's radar and command systems are derived from the S-300 PMU missile, with guidance technology from a *Patriot* missile said to have been sold to China by Israel. If this new missile's performance is similar to the SA-10 surface-to-air missile, the HQ-9 would be useful against cruise missiles and may at least have a potential limited anti-tactical ballistic missile (ATBM) capability.

Recently, the Chinese have embarked on an aggressive programme to procure state-of-the-art Russian SAM systems and related technologies. To date, limited numbers of the SA-101b, the SA-10c, and SA-15 SAMs have been sold to China. These Russian air defence systems are being used operationally to fill gaps in China's air defence structure. However, these systems provide only a rudimentary, limited defence against aircraft and cruise missiles. In an effort to acquire greater air defence depth, Beijing can be expected to make attempts to glean technology from its purchased items to build its own air defence weapons. In addition, China also can be expected to try development of a viable ATBM and ABM capability by either producing its own weapons or acquiring them from foreign sources. Several SAM systems currently in the PLA inventory are assessed to have a limited capability against cruise missiles. These include the CSA-5 and the Hong Qi-7 (HQ-7), a copy of the French CROTABLE systems. The HQ-9 SAM currently under development is believed modeled after the *US Patriot*. It reportedly is intended to provide long range defence against fixed wing aircraft, as well as against TBMs. According to Chinese media reports, the HQ-16 SAM is a Russian-Chinese co-development project, possibly involving SA-11 technology. In addition, Beijing reportedly is developing a short-range SAM system based on the SA-15.

The Chinese Navy's surface units are ill equipped for air defence, particularly ASCMs. Only a handful of the Navy's approximately 60 destroyers and frigates are equipped with SAMs; the remainder are outfitted with anti-aircraft guns of various calibres. The few existing SAM systems have extremely limited ranges and are useful only for point defence. No long range ship borne SAM systems currently exist in the inventory. China is seeking to address its naval air defence shortcomings through the development of a naval variant of the HQ 9 SAM. The PLAN already has deployed albeit in limited numbers a naval variant of the HQ-7. The SA-N-7 SAM system which will be acquired from Russia as part of the SOVREMENNY destroyer deal is a modern, medium range naval SAM system; however, it will have only a limited capability against cruise missiles.

Low Observable (LO) Technologies

China reportedly embarked on an extensive national effort to understand and develop Low Observable (LO) technology in the 1980s. Indigenous efforts have grown in maturity and understanding so as to allow analysis of foreign capabilities and attempt to duplicate past research. Chinese scientists are said to have an excellent theoretical understanding of LO technology, but apparently lack practical experience that comes with decades of applied research. China appears to have begun multiple programmes to apply basic signature reduction technologies to its fighter aircraft programmes and reportedly is developing new fighter aircraft which will incorporate LO technology.

Laser Weapons

China is believed to have a highly developed electro-optic industry, as well as the ability to field blinding laser weapons, including tactical laser weapons. Beijing offered the ZX-87 neodymium laser blinder for sale at defense exhibition in Manila in 1995 and may be developing an advanced version of this system with improved range and anti-sensor capabilities. Although the ZM-87 is intended for use primarily against ground targets, it could also be used against aircraft. China may be developing improved blinding weapons incorporating automatic targeting and countermeasure resistance; it also reportedly is investigating the feasibility of ship-borne laser weapons for air defense. Future laser systems could emphasise the use of advanced optical techniques for improved target acquisition and pointing and tracking. Lasers with increased power and efficiency also are said to be under development.

An article titled 'Made in China' (*India Today,* 25 June 2001) contends that flash floods which ravaged the valleys of Siang river in Arunachal Pradesh and Sutlej river in Himachal Pradesh during June and August 2000, respectively may have been triggered by Chinese authorities in Tibet. Satellite photographs obtained by Indian Space Research Organisation revealed that these floods were consequent upon the release of excess water that had collected in the water bodies of the river basins in Tibet. The former Chief Minister of Arunachal Pradesh, Gegong Apang, a native of the affected Siang valley, had also raised this issue.

Periodically, the UK based Tibet Information Network has reported that China has plans to change the course of rivers in Tibet by carrying out controlled nuclear explosions. While such reports could be speculative in nature, they do suggest options available to China in times of crisis.

PROJECTIONS IN DEFENCE MODERNISATION

A recent Chinese article has projected the progress of China's Defence modernisation over the last two decades. It has spelt out nine areas of progress.

Downsizing : The watchword is "better troops, improved composition and high efficiency". This involved stabilisation of strength of Chinese Armed Forces (PLA) at around 2.5 million, reform of system of leadership, improving fire power, offensive/defensive capability and maneuverability and rapid reaction, and creation of reserve forces.

Fostering Civil-Military Talent : The PLA Interior Service Regulations stipulate that all units should organise collective civil-military training and hold scientific/cultural classes on Saturdays. Approximately one million personnel have attained educational standards above junior college level through on the job training. About 85 percent of all Servicemen have received training in civil occupations, which will enable them to re-settle in civil trades after release.

Military Education : The focus areas are PLA's combat effectiveness in modern warfare: improving capabilities in joint operations, rapid reaction, electronic warfare and logistics. Training capsules on military science and technology have been evolved. Joint-manship is being promoted with a view to winning wars under hi-tech conditions. The educational level of all ranks has been improved with 58.8 percent of personnel being educated upto junior college level in 1997 against 9 percent in 1978.

Technological Achievements : China is the fifth country in the world, capable of firing submarine launched ballistic missiles. It has successfully developed the Galaxy Super Computer capable of 10 billion calculations per second. The Navy has laid a solid foundation for implementing the strategy of off-shore defence. The Air Force has a rational combination of high, middle and low technology weapon systems and a matching combination of fighters, ground attack aircraft and transports. The Second Artillery is able to launch nuclear counter-attacks in self-defence and conventional attacks in depth, independently or in coordination with other arms/ Services. The Army has developed weapon and logistics systems suited to three-dimensional mobile operations. The PLA can control and utilise the electromagnetic spectrum and has upgraded its electronic information support systems.

Legal System : Mechanisms have been established for enforcement of military law and dispensation of military justice.

Logistics : A 'Military Logistics Base System' has taken shape. It has undertaken 20,000 engineering projects aimed at maintenance/upgradation of military harbours, airports, communication centres and missile launch bases. It also caters for construction of accommodation, water and power supply.

Frontier Guards and Garrisons : The weaponry and habitat of Frontier Guards have been considerably improved. Patrolling aboard helicopters and employment of infra red sensors and radars is now reality for these troops. An inter-service Hong Kong Garrison is in charge of the security of Hong Kong Special Administrative Region.

Civic Action : The Chinese Armed Forces have supported national economic construction through various civic action programmes, disaster relief, scientific and technical research, public works projects and environmental protection.

Foreign Exchange Programme : The PLA has established friendly relations with 100 foreign armed forces. There are 90 Chinese military attachés abroad and over 60 foreign military attaches in China. In the last two decades, over 1300 Chinese military delegations visited 80 countries and hosted about 2100 foreign military delegations.

SOME OBSERVATIONS ON DEFENCE MODERNISATION

There are widely divergent views on the increase of Chinese military capability after almost two decades of defence modernisation. This dichotomy is largely due to lack of hard evidence, aggravated by a Chinese tendency to conceal both strengths and weaknesses. China watchers also tend to develop mindsets and get carried away by anecdotal knowledge, sometimes provided by Chinese interlocutors that may be impossible to confirm or refute. American researchers also tend to compare the capabilities of PLA with the US Armed Forces, resulting in biased assessments. Resultantly, conclusions about China's defence modernisation often leave considerable room for varied interpretation. The Chinese Armed Forces have come a long way in the 52 years of their post- 'Liberation' existence. They have gradually given up the ideological linkages with the Chinese Communist Party and are progressively becoming more professional. This trend is expected to continue. There are some interesting and noteworthy trends regarding China's Defence modernisation and capabilities as under:

It is said that China's defence industries have been able to improve their production capabilities and deploy some relatively modern weapon systems. The progress in the field of ballistic missiles is noteworthy. However, the defence industries have not yet been able to realise some of PLA's more ambitious near term modernisation goals. PLA's large holding of out-dated and obsolescent equipment is one of the most significant factors constraining the overall development and pace of its professionalisation. Due to some progress over last ten years or so, it is no longer a PLA of People's War era. But it has a long way to go before it achieves the level of operational capability and technological sophistication its ambitious leadership desires. Given the domestic political stability and sustainable economic growth, it should take another twenty to thirty years to reach that level.

The logistic reforms have tried to address critical weaknesses in China's capability to fight a modern war on its borders. However, they may not have improved China's power projection capability as maintained in certain quarters. The PLA has started focussing on offensive aspects of Information Warfare. They may try to achieve a 'pre-emptive strike' capability against their perceived adversaries' communications, financial and other networks. The developments in conventional missiles and space-based reconnaissance, combined with pre-emptive strategy, could provide China a psychological advantage over its adversaries.

The Chinese Air Force lags behind its modernisation goals. While it may improve its inventory of modern aircraft by 20-25 per cent in five to 10 years, achieving a man-machine mix may take longer. Despite emphasis on naval modernisation, PLA is likely to retain only a coastal defence capability in the foreseeable future.

China's continued growing access to a very broad based foreign military technology from different sources does not assure that the PLA will be able to threaten the Asian balance of power in near future. However, China may double its Defence expenditure in the coming decade. The hardware that China is buying or seeking to buy, indicates its potential future capabilities that cannot be ignored.

China actively seeks to become a strong and modern military power with an ability to promote its national interests within and beyond the region. Its emphasis on lack of transparency, rhetoric and refusal to renounce the use of force tend to remain causes of concern among its neighbours in the region. With regard to India, a (April 2001) RAND study on 'Changing Political-Military Environment' has analysed the India and China conventional and nuclear forces equation. The study reveals that India's conventional forces enjoy a comfortable superiority over their Chinese counterparts in the Himalayan theatre. The Indian Army has superior firepower, better-trained soldiers, carefully prepared defences and more reliable logistics. Similarly, the Indian Air Force has better aircraft, superior pilots, and excellent infrastructure and would most likely gain tactical superiority over the battlefield in a short time in the unlikely event of renewed India-China hostilities. While the Indian Navy is not relevant to any border conflict, it is superior to the Chinese Navy in technology, training, and war-fighting proficiency. The Indian Navy would have little difficulty in enforcing effective surface and subsurface barrier control should any Chinese naval units seek to break out into and operate within the Andaman Sea. China currently has an overwhelming, uncontestable nuclear and missile superiority over India. This superiority is attenuated by two simple realities. Firstly, the political disputes between China and India are too small to warrant recourse to nuclear weaponry on either side. Secondly, the development of India's nuclear deterrent will provide it with a modest means of deterring all but the most extreme Chinese threats.

However, many grey areas still remain in the aforementioned analysis. The Chinese Defence industrial base is more developed than its Indian counterpart and they have a captive market for their low-end technology products. The Chinese are also ambiguous regarding use of low yield or tactical nuclear weapons, or 'clean nukes', within their own territory. Nuclear as well as conventional missiles and its related system is one area where they are undisputedly far ahead of India. Present trends indicate that while the gap between the Russian and Chinese capabilities is considerably decreasing, the Sino-Indian nuclear asymmetry is set to widen. Overall, China's Defence modernisation will lead to progressively augmenting military capabilities. Whilst the intent to use these capabilities through a military option is a different ball game, the widening of the India-China military equation is an important issue which needs to be observed intently, studied in depth and addressed in time.

Chapter 6

PLA in Decision Making

In the framing of China's national policies and related decisions, four inter-related sub-arenas are relevant: strategic objectives, foreign policy, defence policy and intelligence estimates. The role of PLA in these factors varies from near total control over defence policy to limited influence over foreign policy. This aspect is elaborated in the succeeding paragraphs.

Policy Formulation Process

China today does not have a charismatic leader of the stature of late Chairman Mao or Deng. It faces increasingly complex domestic and international problems. These necessitate greater consultation, coordination and agreement among the Party, bureaucrats and senior military leaders in the framing/revision of national strategic policy objectives and the decisions related to them. The closed system and top down approach in tackling most problems/issues compound such problems. Although the avenues of PLA influence over the sub-arena of National Strategic Objectives are relatively few, its role in shaping them is significant and increasing. On foreign policy issues, the senior Party leader usually adopts complex nuances in their interactions with military leadership. These are aimed at retaining the initiative and flexibility by placating, resisting or diluting military views. The process involves personal persuasion, balancing of bureaucratic interests and direct control over the policy-framing channel. Certain key foreign policy areas, like those involving territorial issues or Taiwan's re-unification, may remain subject to military advice. It is generally believed that the military commanders exacted a price in return for their loyalty to Jiang, a larger say in foreign-policy making. The impetus for increased tensions in the Taiwan Strait in 1995-1996, which included mainland missiles being fired into the strait and a series of menacing military exercises, is believed to have come from PLA commanders. Senior PLA leaders interact with the political and bureaucratic hierarchy in a consultative manner. On defence and security related issues, the PLA has a major role and its submissions comprise the clinching factor before a decision is arrived at and a policy framed. The PLA hierarchy has so far ensured that it is not left out of the security policy framing process. It has been observed that the role of PLA in providing strategic analyses and intelligence estimates to the political hierarchy is increasing. This aspect has enabled the PLA to remain in the reckoning when National Strategic Objectives are framed or revised.

The actual functioning of China's national policy framing apparatus is not very integrated, systematic or highly formalised. The personality and influence of the person who heads it often determine the level of influence in the policy/ decision making sub system. All four sub-arenas of the system are inter dependent. It is understood that the weakest link in this process is that between foreign and defence policies. As China opens up and becomes more integrated with the international system, the level of PLA's assertiveness over senior decision making levels will depend on the outlook, relationships and intentions of the emerging political and military leadership. Intelligence estimates and strategic analyses play a major role in China's policy and decision making process than is the case in a democratic setup. Interestingly, China is the one country which reserves 24 seats in its 344 Central Committee for the People's Liberation Army (PLA) officers and men.

Military Diplomacy

In China, military diplomacy plays an important role in pursuance of its national objectives. Foreign relations is not an exclusive domain of the Ministry of External Affairs. In fact in such issues which have important international security linkages like Taiwan, the PLA has a much greater say in policy formulation than the Department of Foreign Affairs. The aim of PLA's military diplomacy is to:

(a) Shape the international security environment in support of Chinese national security objectives.

(b) Improve political and military relations with foreign countries.

(c) Enhance modernisation of Armed Forces and defence industry.

(d) Provide military assistance to friendly developing countries.

(e) Acquire knowledge of modern doctrines, operations, training, military medicine, administration and related non-combat areas.

The PLA seeks to accomplish these goals through its Military Attachés abroad. It also uses an elaborate system of high-level functional exchanges which include ship visits, international conferences, educational exchanges, language training, visiting fellow programme, discussions on operational logistics and training issues, technical cooperation, arms sales, military assistance programmes, arms control negotiations, and peacekeeping operations. The PLA has tailored its foreign relations programme to maximise benefits for itself, while minimising the amount of information it provides to other countries.

One of the most important aspects of the PLA's foreign relations programme is educating the next generation of military leaders. Each year, the PLA's exchange programme sends approximately 500 officers abroad, providing them with firsthand knowledge of the differences between the PLA and other militaries. It has been opined that the most effective way to ascertain China's military and defence policies is to have face to-face contact at multiple levels with Chinese military personnel over a long period of time. Hence, it may be advisable to have more military exchanges with China, on the lines of the Sino-US Annual Defense Consultative Talks, which were initiated in December 1997. It may also be advisable to continue to work on confidence building measures with the PLA, building on successes like the Peace and Tranquility Accord (1993) and the Confidence Building Measures Agreement (1996).

The most visible are the military exchanges, high-level visits, arms purchases, ship visits and functional-level exchanges. The PLA's top six officers average one trip each abroad per year, encompassing 20-25 countries, and host 50-75 foreign counterparts. The Commander and Political Commissar of the PLA Navy and PLA Air Force also take one trip abroad and host from three to six counterparts annually. Functional delegations from throughout the PLA's General Headquarters, Services, and Military Regions are also involved in such military exchanges. During the 1990s, the Chinese Navy paid port calls on over twenty countries and hosted port calls from seventeen countries. These visits have foreign policy value for China, while building the confidence and sea going experience for Chinese Navy personnel. These naval voyages are very different from carrying out major, sustained naval deployments, and should not be interpreted as major steps toward Chinese development of a blue water navy. However, the naval presence outside Chinese territorial waters provides the perception of an expanding naval capability. While media attention tends to highlight the PLA's high-level visits and arms purchases, there are also numerous unreported functional-level delegations that focus on core military capabilities. Functional-level exchanges concentrate on management, policy, structure, training, logistics, operations, and personnel issues for active and reserve forces. As a result of the PLA's massive downsizing over the past decade, Army functional-level delegations have been particularly interested in ideas that can be used to improve the welfare of soldiers and their families. Over the past several years, the PLA has been able to apply the lessons learned in these areas from experimenting with combined arms operations to changing the length of service time for conscription and establishing a non-commissioned officer corps.

Prospects

Many Chinese thinkers have been suggesting that China should have an organisation like the American National Security Council for achieving better coordination. However, this idea has not found favour with the political hierarchy. As China's economy opens up, controlled political liberalisation 'with Chinese characteristics' will come about. This could lead to the PLA assuming more of an advisory role, even in the sub-arena of national security policy. While the PLA may resist this dispensation, it may not be able to prevent it entirely. Notwithstanding the aforesaid, the PLA is likely to remain a major actor in China's transition towards a more open society. As China focuses on increasing its "Comprehensive National Power" in the 21st Century, there are differing views on the scope and pace of the PLA's expanded military relations programme especially as it relates to the PLA's ambitious modernisation plans. Whereas China views its military foreign relations programme as a means to develop friendly relations with foreign countries and modernise the military, some countries view this programme as a means to acquire military hardware and technology, to expand China's naval presence, and to psychologically intimidate its neighbours.

Chapter 7

Chinese Intelligence System

The well organised practice of espionage in China dates back to the 5th Century BC when Sun Tzu's 'The Art of War' emphasised the importance of accurate and timely intelligence as an integral part of statecraft. Subsequently, many Chinese military historians have dwelt upon the collection of diplomatic and military intelligence. In modern history, Chinese intelligence apparatus has operated outside the focus of mainstream counter intelligence efforts. Resultantly, China's competitors are unaware of Beijing's innocuous intelligence efforts directed at them. China's intelligence services play a greater role in supporting policy objectives by targeting or exploiting the technological, economic, political and military infrastructures. While these aspects of Chinese intelligence are known, others like activities of the Dalai Lama's Secretariat and Tibetan refugees in India often remain in the sidelines.

China's intelligence activities are believed to be focussed on factors directly affecting internal stability, regional security, technological and economic development. Its intelligence resources focus on CARs, India, Vietnam and Muslim states wherefrom fundamentalism in Xinjiang receives impetus. American, Japanese, South Korean and Taiwanese military activities along China's Eastern seaboard are also of great interest. Chinese intelligence activities also focus on acquisition of foreign high technology (for dual use), identifying and influencing foreign policy trends (like bilateral policy and trade issues), and monitoring dissident groups (like pro-democracy elements and Taiwanese leaders).

In reality hardly anything is known about the Chinese intelligence agencies and their *modus operandi*. The way the Chinese targeted the Western technology particularly in the field of nuclear weapons and missiles is rather interesting. It is believed that design data with regard to all the seven types of sophisticated US nuclear weapons has been obtained by China over the years through espionage. Millions of computer codes extremely useful for computer simulation and design of new warheads leaked from Los Alamos laboratories to China in 1994-1995. Chinese have an extensive spying network. In February 1986, a CIA employee Wu Taijin was convicted of espionage. He worked in CIA for 35 years, all the while providing classified information to China. During the investigations it was discovered that he was a Deputy Bureau Chief of China's Ministry of Public Security. Some of the Chinese intelligence agencies establish trading companies abroad as a cover to engage in long term intelligence gathering activities. China has more than 3000 companies registered in the USA.

Intelligence Organisations

Ministry of State Security (MSS) : This is China's premier civilian intelligence agency. It was raised in 1983 by combining the espionage, counter-intelligence and security elements of the Ministry of Public Security and the Party's Investigation Department. A Minister of State Security, assisted by Deputy Ministers, heads it.

Organisation: The broad organisation of MSS is given on page 127. The MSS conducts foreign intelligence operations through operatives in the guise of businessmen and diplomats, trade and industry figures, commercial officers, military attachés, journalists, scientists and students. Illegal

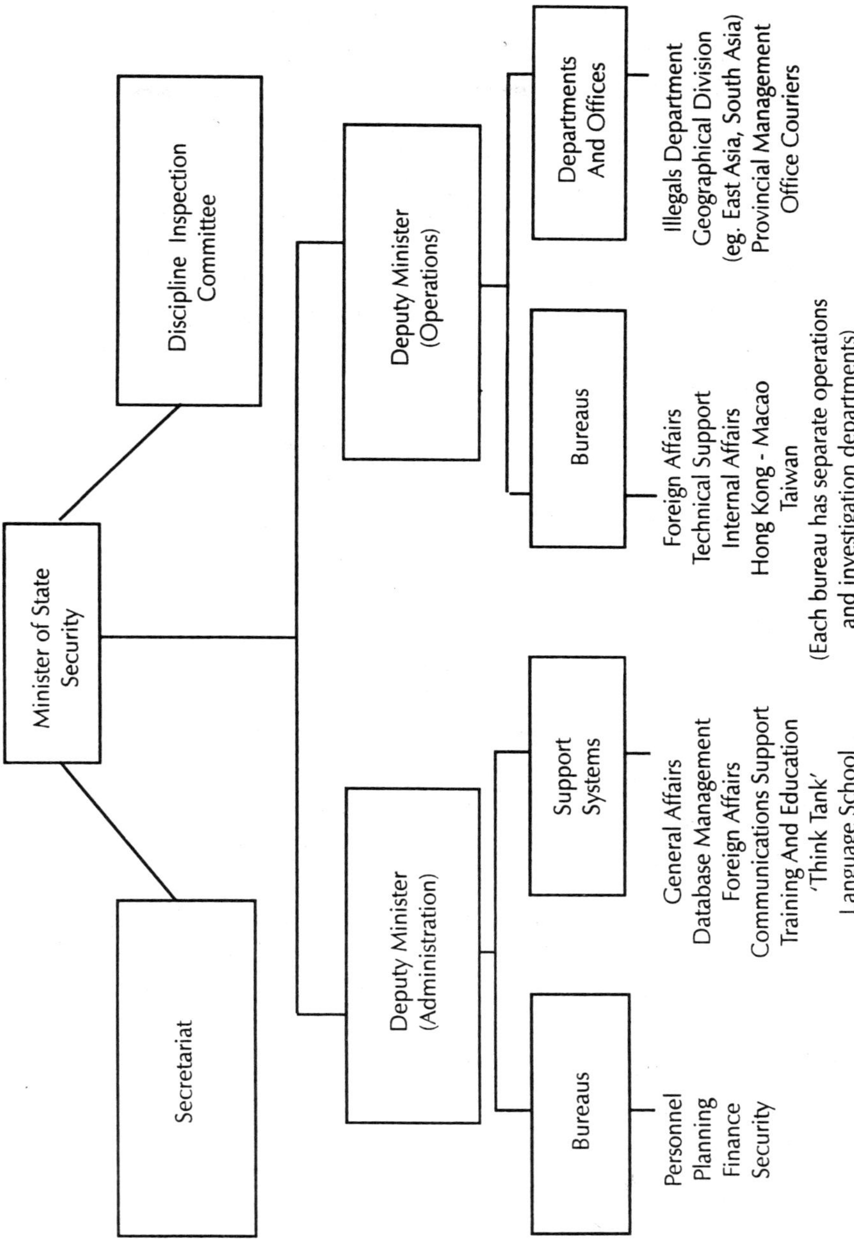

Minister of State Security
Discipline Inspection Committee
Secretariat
Deputy Minister (Operations)
Deputy Minister (Administration)
Departments And Offices
Illegals Department
Geographical Division
(eg. East Asia, South Asia)
Provincial Management
Office Couriers
Bureaus
Foreign Affairs
Technical Support
Internal Affairs
Hong Kong - Macao
Taiwan
(Each bureau has separate operations and investigation departments)
Support Systems
General Affairs
Database Management
Foreign Affairs
Communications Support
Training And Education
'Think Tank'
Language School
Bureaus
Personnel
Planning
Finance
Security

covers are also extensively used. The MSS co-opts many Chinese citizens living or traveling overseas. Such personnel are recruited in China and tasked to acquire the required information when they travel abroad. In certain cases, companies that have access to the information sought by Chinese authorities are purchased by Chinese state-run enterprises. At times, front companies purchase high technology equipment outright. Scholar exchange programmes enable the MSS to legally acquire external intelligence. The returning delegates are thoroughly debriefed to elicit information. The Work Unit is the fundamental building block of China's domestic intelligence set up. These units exist in all places of employment, schools and community centres. They maintain records of membership, family and work history, education, record of travel, political and ideological leanings. The use of Work Units, along with domestic information collection by law enforcement and intelligence agencies, enables effective monitoring of activities considered detrimental to Party interests. Some prominent hotels in China make aggressive use of technical surveillance devices to monitor activities of their occupants. Such devices are also installed in selected public places. However, such surveillance is conducted by different intelligence agencies, including Military Intelligence. The MSS also focuses on penetrating foreign intelligence agencies, investigating espionage cases and anti-party activities. Such activities are of particular relevance in areas like Tibet and Xinjiang, which are not adequately integrated into the mainstream. Activities of foreign media men are also kept under surveillance on a high priority.

The Military Intelligence Department (MID)

The MID functions under the General Staff Department in People's Liberation Army Headquarters. It is known as the 'Second Department'. Its HUMINT acquisition focusses on tactical, strategic and technological intelligence. The MID's activities are largely dedicated to border related intelligence. However, in the arena of technology, MID is probably China's premier technical intelligence gathering agency.

MID's organisation is tabulated below:

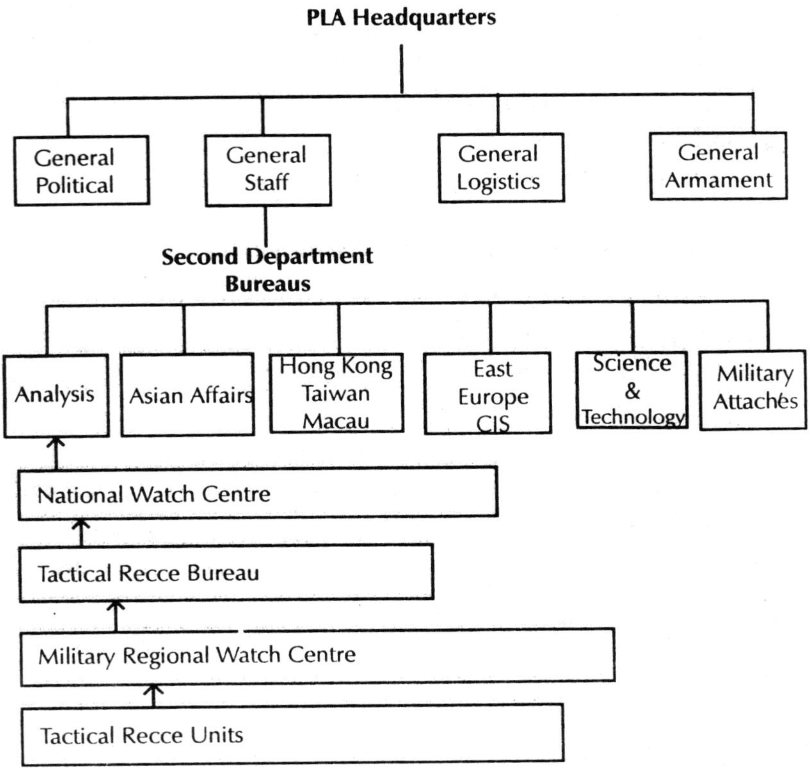

PLA HEADQUARTERS

Departments

(a) *Border Intelligence :* MI units (company at Regiment/ Brigade level) acquire border intelligence overtly and clandestinely. These units have a limited role of covert intelligence collection and propaganda. Coordination is done by the Political Departments at various levels.

(b) *Strategic Intelligence :* The Second Bureau exploits various national assets to conduct strategic intelligence on foreign armies, military technology, identifying foreign markets for Chinese arms exports and conducting covert action campaigns. The military attachés are important constituents of strategic intelligence collection.

Other Intelligence Organisations

(a) *Liaison Department :* The Liaison Department of General Political Department is responsible for studying Sino-Taiwan relations, coordinating propaganda and carring out interrogations.

(b) *Commission of Science, Technology and Industry (COSTIND).* This agency functions under the Defence Ministry and acquires technical intelligence through scholars/ scientists exchange programmes.

(c) *Investigation Department :* The Investigation Department of Chinese Communist Party conducts political investigations, particularly against dissidents.

(d) *Xinhua :* The Xinhua is China's official news agency. Its journalists and staff are involved in overt intelligence collection.

(e) *United Front Works Department :* It is involved in maintaining links with political parties abroad in support of China's foreign policy objectives.

(f) *Political Legal Leading Group (PLLG) :* The PLLG is an organ of the Party with a supervisory role over Chinese intelligence community.

In essence, Chinese have a very effective intelligence set up which has been upgraded from time to time. However even in Western circles, very little is known about its modus operandi.

Chapter 8

Information Warfare and Special Operations

It is believed that India is a global IT power in making. Undoubtedly our software exports and a large involvement of Indian companies in this field particularly in America and some of the European countries have earned valuable foreign exchange for us.

It is generally argued that the overall decline in the Indian exports will be suitably compensated with growing IT exports which has tremendous potential in the coming decade. It needs to be understood that even in this field though we have done well, China has done better, as is evident from the following facts.

	India	China
(a) Number of Computers (in millions)	5.5	16.3
(b) Number of Internet Users (in millions)	6.0	26.5
(c) Number of Cell phone Users (in millions)	5.0	57.0
(d) Fixed Phone lines (in millions)	34.0	164.0

(e) In 1999-2000 China's IT exports were two and a half times that of India.

Interestingly, China has more than double the number of TV sets than India. By 2007, the Chinese language is supposed to become the most dominant language on the Internet. For creating higher employability, India's emphasis should shift over a period of time from software to hardware design and manufacturing. The future lies there. According to Wipro Chairman Azim Premji, if India is to achieve parity with China in hardware, its domestic hardware industry should generate 6 billion USD and not 2.5 billion as at present. The country's PC sales would be 3.6 million and not 1.7 million. All this is a tall order to achieve. The Chinese mad rush for learning English language through crash courses will minimise the advantage which India enjoys in this field at present. IT is a double use frontier technology having high applicability both in civil and military fields. After achieving a high level of proficiency in civilian field, China exploited its applicability to the military to its hilt. It needs to be appreciated that China is utilising contemporary technology to keep abreast with the latest developments in warfare including Information Warfare (IW).

PLA and Information Warfare (IW)

Chinese military strategists who have been influenced profoundly by the thoughts of Sun Tzu have been quick to grasp the impact of the rapid spread in digital communication and information technology and utilise these to enhance their ability to leverage existing military assets. In August 1999, China tested its latest long-range ground-to-ground missile, the Dong Feng-31 that belongs to a class of high-altitude strategic ballistic missile which uses a three-level rocket engine capable of being loaded with multiple warheads. The launch test amply demonstrated Chinese space and ballistic missile capability. The Chinese understanding of IW is fairly comprehensive. They define Information Warfare as :

> "Combat operations in a hi-tech battlefield environment in which both sides use information-technology means equipment or systems in a rivalry over the power to obtain, control and use information. IW is a combat aimed at seizing the

battlefield initiative; with digitized units as its essential combat force; the seizure, control and use of information as its main substance; and all sorts of information weaponry (smart weapons) and systems as its major means. Information warfare is fire assault, operational command for information acquisition, anti-acquisition; neutralisation and anti-neutralization; and for the destruction and anti-destruction of information and information sources."

The PLA's attempts at incorporating IW into its doctrine and operations can be broadly classified into three categories viz. IW activities, psychological operations and deceptive techniques/economic espionage.

The Chinese Information Warfare Centre was set up in 1996 and its primary mission comprises offensive computer attacks by viruses to destroy enemy computers, logistics systems, electric power and water supplies. In fact in October 1997, the PLA began its venture of simulating computer virus attacks in military exercises. Since then, Beijing has been going full speed ahead to acquire an information warfare capability. China has a massive Computer Warfare programme. In April 2001, during the US spy plane crisis, China successfully defaced American web sites during the 11 days stand off. Something similar had happened when the Chinese embassy was bombed in Belgrade. The Chinese swamped the American websites with so much e-mail or 'e-mail bombs' that they were overloaded and knocked out of use. China perceives and appreciates the sensitivity and deep vulnerability of its own information systems. Hence it has accorded the highest priority to defensive Information Warfare programmes and the development of indigenous information technology. In the recent past, the Communication Command Academy in Wshan has emerged as the major PLA centre for IW research. In

December 1998, the Academy established the first IW simulation experiment centre. A task force of 20 PLA IW experts from the Academy wrote the books *Command and Control in Information Warfare* and *Technology in Information Warfare*. The Academy offers 31 command and control related courses with IW at the core. While the PLA's theoretical research on IW is fairly mature, its doctrine and operational field experience is far behind the Western armed forces, but way ahead of India. Recent pronouncements by leading Chinese scientists throw light on the evolving methodology of an IW doctrine. An article by Chang Mengxiong, who serves on the Committee of Science, Technology and Industry of the System Engineering Institute, suggests that military battles during the 21st Century would unfold around the use of information for political and military goals. Accordingly weapons, soldiers and combat platforms would be information intensified. Weapon systems would be able to acquire and use information provided by the targets themselves to correct trajectory and over-the-horizon precision attacks would become the primary form of firepower attacks. Soldiers would be equipped with multimedia electronic information equipment comprising personal radio communication nodes, Global Positioning System (GPS) nodes, a computer and network function. Night vision devices, identification (friend or foe) appliances, clothes with adjustable temperature and colour etc. would enable these information-intensified soldiers to receive highly concentrated commands and information about combat. While the actual acquisition of such capabilities for field force will take a long time, the PLA thinking and conceptualisation on IW appears to be far ahead of all the armed forces in Asia with the probable exception of Japan.

China is aiming to use space technology to monitor its adversary's missiles launching capabilities and strategic weapons systems over land and sea. This field still remains unexploited to a great extent but Chinese strategists see a distinct possibility of a space weapon system as a third weapons revolution after conventional and nuclear weapons. They are fairly certain that a new generation of space weapons will overtake nuclear weapons as principal strategic weapons in the first half of 21st Century.

The PLA Navy meanwhile recognised the fact that in the 21st Century cutting edge technologies are most likely to be applied first to naval warfare and hence there is a need to recast the concepts of naval operations to keep up with the technological revolution. There are efforts on to integrate the various components of land-sea, land-air, surface-undersea and sea-air warfare with the electromagnetic spectrum to extend the scope of naval operations. It is anticipated that the integration would be achieved by a cohesive combination of spacecraft, military satellites, space shuttles and permanent space-based platforms that would observe and control maritime operations. Concurrently, the PLA Navy is also actively pursuing research in a variety of state-of-the-art technologies to speed up the development of naval weaponry. Chief among them are microelectronic technology, stealth technology (to make ships and cruise missile stealth capable), infrared technology (both offensive and defensive) and precision-guidance technology. Certain doctrinal modifications are also being pursued by the PLA Navy to enhance the combat effectiveness of its forces. Thus, the Navy is expected to replace concentration of force with concentration of firepower in view of the increased communication options availability, which in turn would reduce battle casualties. Submarines due to its inherent covertness and power would increasingly be used to carry out underwater raids as well as to deter opponents. The concept of considering the enemy's information network as an important target has also gained significance owing to the vulnerability of these targets. There is a subtle move to integrate command communications among the armed forces as perhaps the first tentative steps towards a joint action strategy.

Space Missions

Besides its own efforts to launch a possible first manned space mission by 2003, China has been discussing joint space missions with European Space Agency including Spacecraft mission 'Double Star'. The two Spacecrafts could be launched in 2002 and 2003 designed to collect data about the pattern of magnetic field and the behaviour of electrified particles in different layers of magnetosphere. It is believed that China is building a space telescope that could rival the Hubble telescope–a five year project planned to be completed in 2005.

Moscow and Beijing currently have 11 joint space projects underway. The Chinese have also shown strong interest in Canadian and German radar satellite capabilities. They may go for the purchase of synthetic aperture radar satellite systems. Since 1998, China and the United States have signed three inter-governmental agreements on launching services. Many US satellite manufacturing companies have signed agreements on commercial satellite launching services involving some 30 satellites. The satellite 'Sinosat', jointly developed by China, France and Germany, was successfully launched in 1999. In the field of satellite development, launching and research, Beijing has promoted technical and commercial co-operation with 70 countries including the UK, Brazil, Italy and Pakistan.

Psyops and the PLA

Apart from the use of normal conventional means to obtain and deny information, the Chinese view of IW extends its definition to encompass the psychological aspects of motivating people and influencing the populace by extensive employment of electronic media and leaflets. Current Chinese power projection capabilities are inadequate as can be gauged from the muted response to Taiwan when the Taiwanese Premier Lee Tung-Hui stirred the age-old hornet's nest by raising the issue of state-to-state relationship between his country and China. However, Chinese attempts at psychological warfare continue unabated. Recent publications have shown that China is interested in using psychological warfare to supplement traditional strategic approaches, exploit the power of modern communications technology and support national policy objectives across diplomatic, economic and military spheres. Both traditional and modern Chinese strategists have been of the view that "it is better to attack the enemy's mind than to attack his fortified cities". This is considered the traditional origin of psychological warfare. But achieving victory without combat requires careful planning, a prosperous and contented populace that fully supports the state, well-trained and highly disciplined troops and absolute secrecy within alliances and during operations. The importance of psychological operations in the Chinese mindset can be gauged from the description of psychological warfare by eminent military strategists as the "fourth kind of warfare", after ground, sea and air warfare. Accordingly this form of warfare will play a major role in China's strategic policy. The four basic objectives of psychological warfare as envisaged by the Chinese are :

- To attack and influence the target's policy makers, causing their position to waver and make them choose a policy that is in fact disadvantageous to themselves.

- To instill fear and awe in the minds of enemy soldiers, destroy their mental equilibrium and boost the morale of one's own troops.
- To create panic among the adversary's civilian population by inducing fear of war, war weariness and antiwar feelings.
- To obtain the sympathy and support of neutral states, solidify coalitions among allied states and strengthen the desire for victory in one's own country.

There are efforts to carry out psychological warfare across a broad spectrum of national activities, including strategy, military technology, politics, diplomacy, religion, economics and propaganda to demonstrate national strength. In addition, increased military strength and technological capabilities that are required to support psychological warfare will serve as deterrents.

Modern communications technology, which keeps the battlefield commander informed of major political and diplomatic developments and military movements, is also being harnessed to raise the influential power and deceptiveness of psychological warfare operations. Greater speed and more destructive precision weapons enhance psychological warfare's deterrence and coercive value. The PLA is developing a psychological warfare doctrine by adopting enhanced training and carrying out supporting research to build up an elite cadre of specially trained personnel and psychological warfare units assigned to the group armies and divisions. Using new communications technology, they would be linked by a psychological warfare command system network. The PLA routinely conducts enhanced training programmes in which troops incorporate psychological warfare operations during live-fire exercises.

Deception and Economic Espionage

The Chinese have always been traditionally masters at deception. Current Chinese military leadership has successfully put deceptive techniques to work by extensively employing 'dual-use' technology originated in the United States for purposes which it was not intended for. A classic example is the sale of machine tools from a former B-1 bomber plant by McDonnel Douglas to the China Aero-Technology Import-Export Corporation. The technology was ostensibly bought for manufacturing civilian aircraft but actually put to use in a military production facility. China's information warfare centre also indulges in economic and industrial espionage. The Cox Committee Report that details China's successful attempts to steal the US nuclear secrets by gaining access to the US Energy Department computers is a case in point. Another area of information operations undertaken by the Chinese, which is particularly disconcerting to India, is the investment in chip production. Doctored chips and software that can plant viruses in information systems are rapidly emerging from Chinese markets. These electronic components that are extraordinarily low priced make it a lucrative buy for the countries in Asia including India with the inherent element of risk.

Conclusion

Chinese national strategy seeks to build "comprehensive national strength" by enhancing China's international standing, improving weapons and telecommunications and promoting continued economic development, while maintaining internal stability. Traditionally, compared to the armed forces of industrially developed countries, the PLA had always studied tactics from a strategic rather than the technological angle. This was chiefly due to knowledge limitations and scientific backwardness of PLA. Of late, however, there has been a perceptible shift in the Chinese military leadership and character towards incorporating the newest forms of warfare into their current concepts. Further, the easy accessibility of information technologies and advanced military hardware in the international arms market is sure to propel Chinese talent to exploit this availability to upgrade its military capabilities. According to a report, China has developed the state-of-the-art laser defence technology that can intercept and shoot down low flying missiles. Exploitation of current technologies by the PLA is bringing about profound changes in the operational concept of warfare. Use of satellites, high altitude aircraft, helicopters, unmanned flying equipment, sensors and digital communication are all envisaged in future warfare. Technological breakthroughs by the PLA beginning with China's entry into the nuclear club in 1964, followed by its first space satellite in 1970 and its first full range intercontinental ballistic missile in 1980 has demonstrated Beijing's strength to leapfrog obstacles in weaponry development. It is, therefore, fairly certain that China would be a force to reckon with in this newest form of warfare, at least in Asia by about 2020 or so. The Indian top brass need to take a serious note of it.

Chapter 9

Defence Industries

At the time of its 'liberation' most of China's defence industries were sabotaged by the Nationalists. By early 1950s, China managed to revive some ordnance, aircraft and ship-building, and communications industries. By 1959, China had established or revived 120 defence industries, with extensive Soviet assistance. These industries had a work force of approximately 700,000. China's defence industries had a major setback with the withdrawal of 11,000 Soviet scientists and technicians in the mid-1960s. Another negative aspect was the subsequent fallout of the Cultural Revolution. After the Mao era, the Commission of Science, Technology & Industry for National Defence (COSTIND) was set up in 1982 as the leading agency for all defence-related research, development and production. The progressive opening up and modernisation drive during 1980s-90s gave considerable impetus to Chinese defence industries. It also boosted Chinese production of low-tech weaponry, which was liberally exported to developing Asian and African countries, on 'friendship prices'. China accrued considerable strategic and some financial gains from such arms export policy. In 1997 Party Congress, a radical agenda was laid out for re-structuring and down-sizing of the Chinese defence industries and opening them up to free market forces. A three-year time table was laid out for implementation of these reforms.

Re-organisation

The charter of COSTIND had been progressively enlarged to include management of the process of research, development, testing and production in Chinese defence industry. In the 1990s, it was re-organised with the military component being designated as the General Armament Department (GAD) and the civilian component was separated from it. The GAD now oversees defence procurements and weapon programmes. The 'civilian only' COSTIND oversees military oriented scientific/technological research, development, production and formulates/ implements defence industry reforms.

The COSTIND's earlier five pseudo-defence corporations were restructured into following 10 defence industry (enterprise) groups as tabulated:

Old Corporate Entity	New Enterprise Group	Major Products
Aviation Industries of China (AVIC)	China Aviation Industry Corporation I	Fighter aircraft, bombers, transports, advanced trainers, commercial airliners,
	China Aviation Industry Corporation II	Helicopters, attack aircraft, light trainers, UAVs
China Aerospace Corporation (CASC)	China Aerospace Machinery & Electronics Corporation I	Space launch helicopters, satellites, missiles
	China Aerospace Machinery & Electronics Corporation II	Missiles, electronics, other equipment
China Ordnance Industry	China North Industries Group	Tanks, armored vehicles, artillery

Old Corporate Entity	New Enterprise Group	Major Products
Corporation (COIC)/ NORINCO	Corporation	Ordnance, Infantry Weapons
	China South Industries Group Corporation	Miscellaneous ordnance, automobiles, motorcycles
China State Shipbuilding Corporation (CSSC)	China State Shipbuilding Corporation (Southern shipyards, based in Shanghai)	Frigates, smaller surface combatants, commercial ships
	China State Shipbuilding Industry Corporation (Northern shipyards, based in Dalian)	Destroyers, commercial ships
China National Nuclear Corporation (CNNC)	China National Nuclear Corporation	Nuclear energy development, nuclear fuel and equipment
	China Nuclear Engineering & Construction Group Corporation	Construction of nuclear power plants, other heavy construction

These reforms were expected to lead to large trans-sectoral and competitive defence conglomerates on the lines of integrated defence companies like British Aerospace. However, this does not appear to have happened on ground because of the following reasons :

(a) Rationalisation has been much slower than planned.

(b) Over-staffing.

(c) The new entities are still running in losses and continue to be heavily subsidised.

(d) Technology absorption remains a problem area.

China's defence industries have achieved some success in technology leveraging. As a result, a few of these industries have improved production capabilities to develop and deploy relatively modern weapon systems :

(a) China has made real progress in the areas of ballistic and cruise missiles, and, to a lesser extent, surface-to-air missiles.

(b) The shipbuilding industry has had some success in leveraging 'soft' commercial technologies like management techniques.

(c) The acquisition of dual use technologies in telecommunications, information processing, fibre optics, cellular/wireless communications and Internet has benefited the PLA.

(d) China is progressing in the development of a fourth-generation fighter aircraft, despite delays.

(e) The advice rendered by Russian and Israeli advisors/technicians could assist Chinese in areas of design, engineering and systems integration.

Present Status and Prospects

Overall, however, the accomplishments of Chinese defence industries tend to be the exceptions rather than the rule. Particularly when it comes to fighter aircraft, various ships and ground equipment, the Chinese still confront severe problems when moving from prototypes to production runs. China's defence industries fail mainly due to endemic technical, structural, and cultural problems. A shortage of technical skills, over capacity, under funding, and a bureaucratic, hierarchical and risk-averse work environment have all impeded technology leveraging. Moreover, current defence industry reforms do little to remove or overcome these impediments. Western experts have opined that PLA will find it difficult to rely on its domestic defence industry to meet many of its more ambitious near-term military modernisation goals.

Chapter 10

PLA's Logistics Support System

The PLA was evolved from a large manpower oriented guerrilla force based on the concept of living off the land. In the initial stages, no serious thought was given to improvement of logistics infrastructure. As PLA's doctrines graduated from "People's War" through "Local War to Limited War Under High Technology Conditions", the need for reforming the logistics system was increasingly felt. In mid 1993, the Chinese sent a high-powered military delegation to the USA, UK, Malaysia, Pakistan and India. This delegation spent about 10 days in India and visited our depots, and administrative and training establishments. It was after extensive research and spadework that PLA initiated logistics reforms. The PLA has implemented many reforms since 1998, some of these are intended to improve its ability to conduct mobile warfare.

The PLA's major vulnerability is its inability to sustain large forces in intense, fast-moving combat operations. A 10-year programme to restructure the logistics system was announced in 1998. Based on the same, the Chinese logistics support system for combined operations was organised as explained:

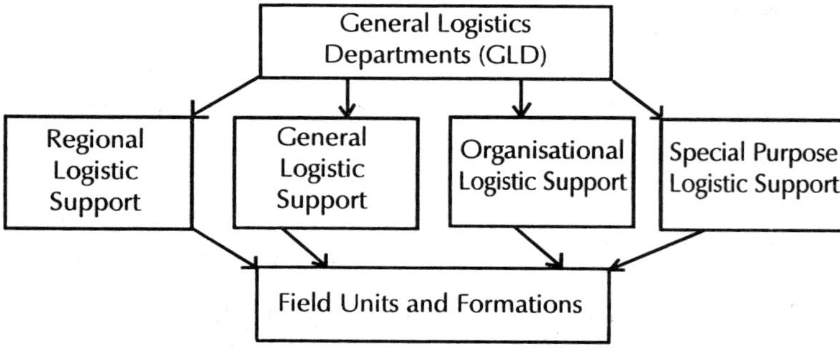

The detailed organisation of **PLA GLD** is as follows :

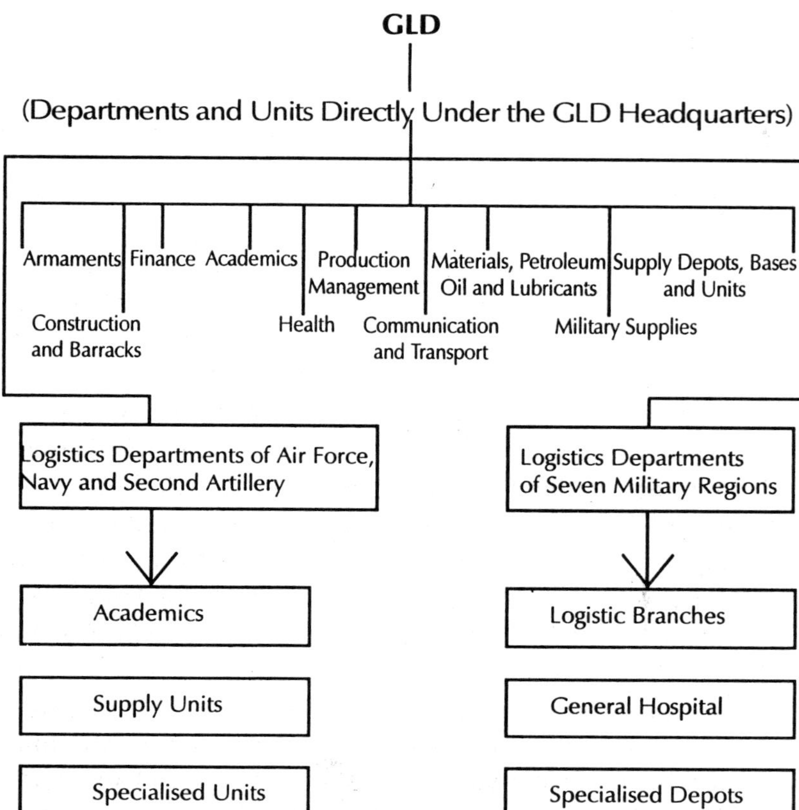

Reforms

An All-Army Joint Logistics Implementation Plan was issued in January 1999, with the following objectives:

(a) Integration of logistics establishments of the three Services.

(b) Standardisation of supply work for units.

(c) Conversion of officers perquisites into cash allowances.

(d) Adoption of 'scientific' management techniques.

(e) Improvement of mobile logistical support for units operating away from their base areas.

At both the national and the local levels, the separate Services of the PLA—Navy, Air Force, and Second Artillery - have maintained separate logistical infrastructures. There is no separate Ground Forces (Army) Headquarters in the PLA. The Military Regions (MRs) exercised nominal control over the other Services' logistics structures in the regions, but actual control centered in the Service Headquarters at Beijing. Under the reform programme, those support functions that are common to all Services are being merged into new Joint Logistics sub-Department (JLD) in each of the MRs logistics department. The Services will retain separate logistics structures for supplies and supra-unique to their own systems. Under the rubric of "network-style zoned supply work," the Services are surrendering hospitals, medical supply and maintenance units, fuel stocks and distribution networks, general supply warehouses, vehicle supply and maintenance units, and some general transportation units to the control of the MR/JLD. The logistics reform programme calls for strengthening the role of the MR and Military Districts (MDs). The reformed logistic system is based on MR and MD and is responsible for joint supply to various elements of the three Services within them. This has ensured enhanced synergy, reduced redundancy and affected economy of effort.

Hospitals, fuel and supply depots now cater to the requirements of all Services within MR.

Socialised logistics is an economic move, seeking to capitalise on the efficiency of the market and reduce the overall cost of support functions. The main thrust is to relieve military units of unnecessary overhead and allow them to focus on operational responsibilities. An open equipment technical support system is considered an effective measure to "contain the army in the people, unite in peace and war". With this path breaking change, it becomes possible to reduce unnecessary storage and maintenance ability in the system and decrease the expenditure of the system. Socialised logistics will contract out most of the day-to-day service functions that sustain the PLA in garrison including fuels supply, vehicle spares, garbage collection, tree planting, childcare and management of officers housing. Outsourcing of support functions has commenced with garrison units and static PLA institutes in major cities. This is to be gradually expanded to include the entire force.

PLA has realised the vulnerabilities of the static, fixed-depots structure of its logistical system. Throughout the 1990s, the Ground Forces (Army) and the Air Force (PLA AF) have designated Rapid Response or 'Fist' units, and they and the Second Artillery have practiced rapid deployment some distance away from their home base for field training exercises. The logistics reform programme seeks to restructure tactical and operational level logistics elements to provide better support for out of MR deployments. This is considered the minimum requirement if the PLA is to develop mobile warfare capability. While retaining some fixed-site logistics structures, PLA is seeking to graft on a mobile structure to provide forward support to the manoeuvre units. The MRs have experimented with a variety of organisational structures to provide mobile support. The current focus is on two types of organisations:

(a) Emergency Support Brigades, consisting of medical battalion, supply battalion, a fuel battalion, transportation battalion and emergency military depots, encompassing ammunition depots, fuel depots, warehouses, field clinics and transport units.

(b) The PLA Navy (PLA N) has experimented with emergency support detachments to repair warships at sea. The PLA AF has experimented with composite mobile support detachments, comprising elements from different organisations, to accompany fighter units deployed in South Eastern coastal belt.

Stock numbers, requisition procedures, inventory standards and other supply operations vary from region to region. This was a major hindrance to the PLA's ability to deploy units across China. To offset this, standardisation became a major priority of the logistics reform programme. It aimed at gradually establishing a set of supply standard systems, with corresponding rules and procedures, which are easy to implement.

Another theme of logistics reform is the conversion of officer perquisites into cash allowances. The transition to a market economy is converting almost all of those exchanges to cash transactions. Military personnel costs have skyrocketed as the PLA has to provide its personnel, especially officers, with ever-increasing volumes of cash to pay for goods and services. The Officer Compensation System will continue under the logistics reform programme, with particular emphasis on housing and vehicles. The medical and pension costs for veterans killed or injured in the line of duty are being shifted to state insurance system. In pursuit of similar objectives is the establishment of new ranks and pay scales for non-commissioned officers in the July 1999 revision of the Soldiers Service Regulations. The Regulations create six new grades, NCO 1 through 6, with a pay scale that increases in line with rank and service. Other changes include:

(a) In the past some Army units carried out commercial activities for raising maintenance funds. This had its attendant problems. Since end 1998, such activities were banned.

(b) 'Skip Echelon Logistics' involve streamlining logistics operations within MR, direct links between the MR Joint Logistics Department and regional support units.

(c) Increasing the standard dietary allowance for soldiers, with more meat and greater variety (but no increase in budgetary allowance for food), improved field kitchens, combat rations of roasted rice and hard tack instead of uncooked rice.

(d) The 1998 flood relief operations provided valuable real-world experience in rapid re-supply. This resulted in focusing on faster inter-modal transfer of material (from land and air, to sea modes).

Trends and Implications

If successful, the logistics reforms will improve the PLA's ability to sustain its forces in high-intensity combat on China's periphery. But they do not contribute significantly to regional or global power projection. However, the following components required to sustain large manoeuvre forces in combat beyond China's periphery remain absent:

(a) Logistics elements organic to manoeuvre units, transportation capability, first and second line maintenance, evacuating and repairing equipment casualties, mobile fuel and supply stockpiles.

(b) Port opening and airfield opening packages, capable of creating the far end of a logistical flow structure.

(c) Greater airlift and sea lift capabilities; combatant escorts and refuelling units to get the lift to its destination.

Presently, only 10 percent of PLA's logistic units are in a position to support Rapid Reaction Forces (RRFs). Available trends point towards increasing attention being paid to technical support and expansion of civilian transport and infrastructure capabilities. On balance, PLA may continue to retain a large logistic base with pockets of excellence in the form of emergency support units to cater for RRFs. In totality, at the present pace of restructuring the logistics system it will not be able to provide effective campaign level support till 2010 or so.

CHINESE ACTIVITIES IN ASIA

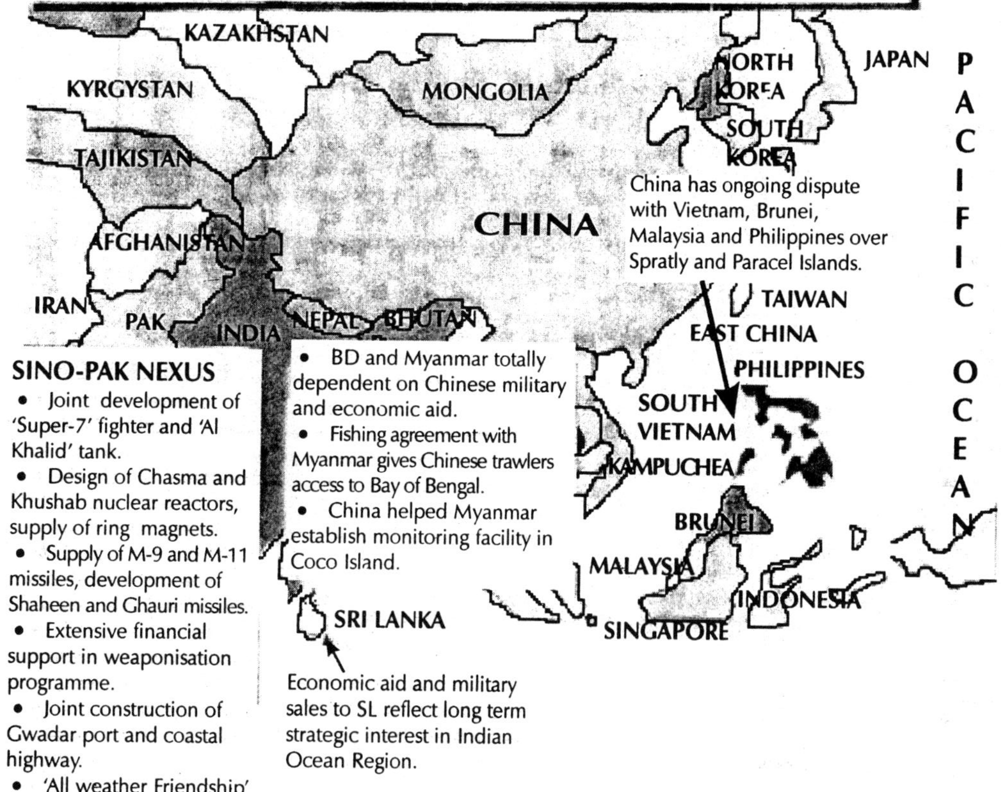

Chapter 11

China's Activities in India's Neighbourhood

The pendulum of global geo-strategic activities is distinctly swinging towards Asia in the 21st Century. Besides the traditional 'hot spots' in West Asia and Asia Pacific, the developments in Afghanistan, Pakistan and Central Asian Republics and renewed American strategic and economic interest in and around Indian subcontinent will have impact on our long term security considerations. The developments in Indonesia, Philippines, Sri Lanka, Bangladesh, Myanmar and Nepal will directly or indirectly impinge on our economic interests and national security. The activities of drug mafia between Golden Crescent and the Golden Triangle have important linkages in the insurgency in our North-East. Further, while the shape of the new world order is yet to clearly emerge in the transition towards multi-polarity, India, China and Japan will play a predominant role in Asian affairs. It is therefore, important to briefly study the trends of the Chinese activities in our neighbourhood and to analyse its impact on India's economic and strategic interests.

Sino-Pak Nexus

The downswing in India-China relations following the 1962 Chinese aggression paved the way for convergence of Sino-Pak relations. This was strengthened with Pak ceding over 5000 sq km territory in Pakistan Occupied Kashmir (POK) to China under an agreement in March 1963. Sino-Pak relations are now an "all weather friendship", and the Karakoram Highway symbolises it. A notable development during Vajpayee's bus diplomacy to Lahore in February 1999 was that the three Service Chiefs absented themselves from meeting him at Lahore, instead choosing to greet the visiting Chinese Defence Minister Gen Chi Haotian.

China has been continuously assisting Pakistan in its nuclear weapon and missile programmes to ensure its parity with India in this field. Numerous Chinese, American and Pakistani press reports have indicated that China has supplied Pakistan with the complete design of a nuclear warhead carried by a ballistic missile in its fourth nuclear test conducted in 1966. The long list of material and equipment given to Pakistan include enriched uranium, tritium for boosting the yield of nuclear devices, ring magnets, a special furnace to melt plutonium or enriched uranium into the shape of a bomb core, M-11 ballistic missiles and a factory to produce ballistic missiles. It also provided Pakistan with an unsafeguarded plutonium reactor at Khushab. China has assisted Pakistan in the development of both the enriched uranium and plutonium routes to nuclear weapons and provided the necessary delivery systems. The Pakistani M-9 and M-11 missiles which target Indian population centres from Delhi to Bangalore are of Chinese origin. A recent RAND study by Dr Ashley Tellis mentions that Pak's nuclear arsenal comprises 20-25 nuclear weapons of Chinese design. Despite assurances to the Clinton and Bush administrations, a report in *Washington Times* (6 August 2001) indicates that the Chinese transferred components of Shaheen I and Shaheen II missiles to Pakistan on 12 occasions during the year 2001. The latest such transfer was detected by a US spy satellite on 1 May 2001. Although the Chinese have denied such reports, the US representatives have reserved their comments. Chinese assistance to Pakistan has been clearly indicated through:

(a) An agreement to develop Gwadar deep waterport and Makran Coastal Highway that was signed in July 2001. China is to provide $ 200 million soft loan towards Phase I of the one billion project.

(b) Joint development of 'Super-7' fighter and 'Al Khalid' main battle tank.

(c) Comprehensive arrangements for supply of weaponry, ammunition and spares to Pak on 'friendship prices'.

(d) Pakistan acts as a conduit for Chinese links with Taliban, to the extent that they made an unexploded US Cruise missile, fired during mid 1999, available to Chinese technicians for inspection.

(e) Extensive intelligence sharing, and moral, diplomatic and material support.

Sino-Myanmar Relations

While India and the rest of the world was busy in the late 1980s condemning and pushing Myanmar's military junta into a corner, China exploited this opportunity by providing it moral and military support. Today, Myanmar is a virtual military ally of Beijing and a captive buyer of Chinese low-tech weaponry and hardware. In fact there are frequent people-to-people contacts between inhabitants of North West Myanmar and China's Yunnan Province. In this part of Myanmar, the Chinese are called 'brothers' and not foreigners, by the local Burmese. The military junta has allowed Chinese military, commercial and political influence to extend through Myanmar into the Bay of Bengal and towards approaches to the Straits of Malacca. China has been building a rail link from Yunnan Province through Myanmar to the Bay of Bengal. Chinese military sales to Yangon even at friendship prices are estimated at over $ 2.5 billion. These involve fighter aircraft, tanks, infantry combat vehicles, artillery, naval vessels and many other items. Chinese military assistance has enabled an increase in Myanmar's Armed Forces from 180,000 to 450,000 troops during the 1990s. Almost 90 per cent of Myanmar's military equipment is of Chinese origin. This includes ammunition, repairs, training and provision of spare parts. There has been significant Chinese investments in Myanmar's transport infrastructure. Chinese funded projects include bridges, dams, roads, waterways and ports. The Chinese are involved in developing the Irawadi river – a project that has ensured the presence of 2,50,000 Chinese of all sorts. The growing Chinese influence in this region is best illustrated by the Sinofication of Mandalay, where even the signboards are in Chinese language. A 1996 Sino-Myanmar Fisheries Agreement gives up to 300 Chinese trawlers access to Myanmar's waters. This, coupled with sale of Chinese riverine and coastal vessels and container ships, has given considerable access to Chinese vessels to Myanmar's ports and waterways. There are periodic newspaper reports of Chinese presence in Coco and Hainggyi Islands off Andaman and Nicobar islands. The Chinese have reportedly assisted in the setting up of monitoring stations there, which could be developed to monitor Indian Naval activity in the Bay of Bengal as also missile tests in our range at Chandipur, Orissa.

Sino-Nepal Relations

The Sino-Nepal ties were formalised after King Mahendra ascended the throne in March 1955. King Mahendra paid his first visit to China in October 1961 when India-China relations were at their lowest ebb. During this visit, the Nepal-China Boundary Treaty was signed. Significantly, during and after the India-China conflict, Nepal never publicly expressed solidarity with India. King Birendra, who ascended the throne in January 1972, continued the talk of 'equal friendship with

India and China'. During his reign, Chinese foreign policy considerably influenced the course of Indo-Nepal relations. During the pro-democracy movement in early 1990s, there were media reports that Nepal had procured weapons from China via the Chinese built Friendship Highway. Besides this prestigious project, extensive road networks has been developed linking Tibet with Nepal making it possible to open a new front against India through Nepal, if required. There were reports that the Chinese had supplied air defence guns to the Royal Nepal Army. Subsequently, Chinese firms have been involved in road construction and development in Nepal, including in the Terai belt bordering Central India. Media reports in the year 2001 indicate that an influential section of the elite in Kathmandu have advocated virtual political alliance with Beijing. Interestingly, when communism is on the decline the world over, the movement is making impressive gains in Nepal. The Communists in Nepal are openly aligned with China.

Sino-Bangladesh Relations

China being a staunch ally of Pakistan had opposed the liberation of Bangladesh during 1971. However, the murder of Sheikh Mujib and subsequent developments saw a warming up of Sino-Bangladesh relations and by mid 1970s, the relations had become very close. Bangladesh's failing economy and China's design to increase its influence amongst the Indian Ocean states provided convergence of Sino-Bangladesh interests. China is the provider of almost three-fourths of the weapons and equipment held by Bangladesh's Armed Forces. Many Bangladeshi officers are trained in China. Chinese infrastructure companies have been awarded many contracts by under-cutting the tenders quoted by others. During mid-2000, China agreed to finance a major Convention Centre for the NAM Summit in Dhaka. Although an India-friendly Awami League Government headed by Sheik Hasina was in power in Bangladesh till last year, it failed to check the growing Chinese influence.

Sino Sri Lanka Relations

Sino-Sri Lankan relations were established during the 1950s. Presently, this relationship has been kept up on the basis of Sri Lanka's requirement of cheap weaponry and defence equipment to fight the continuing insurgency. The Chinese meet most of Sri Lanka's defence requirements. Thus, Sri Lanka's dependence on Chinese weapons is more due to economic compulsions rather than political or ideological reasons. Chinese Prime Minister Zhu Rongji visited Sri Lanka during May 2001, as part of his tour to Pakistan, Maldives, Thailand and Nepal. Keeping in view China's increasing importance for Indian Ocean region, Sino-Sri Lankan relations will continue to receive impetus from the Chinese side.

China, ASEAN and SAARC

China is a dialogue partner of ASEAN and has the privilege of holding Sino-ASEAN summits. Lately, China had even opposed the holding of Indo-ASEAN dialogue . Simultaneously, China is seeking to enter the Mekong-Ganges Corporation Group (India, Kampuchea, Laos, Myanmar, Thailand and Vietnam). It is understood that the delay in holding the Indo-ASEAN summits has been largely due to Chinese opposition to such an arrangement. China has also shown interest in becoming a member of SAARC. Possibly, Pakistan may press its case for membership. These machinations clearly suggest China's efforts to increase its sphere of influence in South East and South Asia, while endeavouring to degrade Indian influence in these regions.

Implications

From the international relations point of view, China's pro-active thrust in its neighbourhood to safeguard its interests may pass off as normal diplomacy. China is a major power in Asia and is fully justified in handling its neighbours the way it likes. However, it has an adverse effect on India. The ring of Chinese activities in India's backyard is to restrict Indian influence in a domain which has historically and traditionally been India's area of influence. China's natural area of influence is primarily East Asia. Its enhanced activities in South Asia is to keep India 'grounded' in a manner that it is not in a position to challenge growing Chinese influence in rest of the Asian region. China's growing presence in Indian Ocean is to progressively ensure India's 'containment' to the maximum possible extent. It is thus being ensured by China that it gets more strategic and economic space even at India's expense. The heavy dependence of South Asian and South East Asian countries on Chinese weapons, spares and training is an important development which India can ill afford to ignore. The Sino-Pak nexus has shown no signs of abating. By propping up Pakistan, China has done little to allay India's concern. The Sino-Pak nexus has led to a progressive increase in the military capability of Pakistan and has emboldened it to continue its proxy war in J&K. This has stretched India's military and other resources and caused considerable weakness in the socio-economic fabric. These developments have served to holistically degrade India's legitimate sphere of influence. This is to India's utter discomfort and may become a source of friction between India and China.

Chapter 12

Developments in Tibet

The original (pre-1949) "Roof of the World" Tibet comprised three administrative regions of Kham, Amdo and U-Tsang. It spanned an area of 2.5 million (mn) square km. However, the present Tibet Autonomous Region (TAR) of China spans 1.2 mn square km and comprises only the former U-Tsang region. Amdo was designated as Qinghai province since 1955 and Kham divided into five administrative prefectures during 1953-1957. Through trifurcation of Tibet, the Chinese have cut Tibet to size, for effective regulatory administrative control and to contain internal dissidence and resistance by Tibetans. The Tibetan Plateau is the highest (average altitude 4000 mtrs) and largest plateau in the world. It is the source of 10 river systems. Of these, three flow into India: Tsang Po as the Siang (later Brahmaputra, into Arunachal and Assam), Langchen Khabab as Sutlej (into Himachal Pradesh) and Senge Khabab as the Indus (into Ladakh).

The Chinese have carried out large-scale population transfers of the majority Han population into the Amdo (Qinghai) and Kham administrative regions. The UK based Tibet Information Network periodically propagates these demographic changes as being aimed at Sinocization of Tibet with a view to reduce the Tibetans to a minority in their own homeland. Hans, mostly ex-servicemen, constitute 65 percent of population at Lhasa. They are also encouraged to settle in other cities in Tibet. However, the Chinese openly refute such allegations. Through such manipulations, the Chinese have largely strengthened their hold on Tibet and confined dissidence to the area of 'Tibet Autonomous Region', which is only 50 percent of 'Greater Tibet'. Numerous measures are being taken to reduce the influence of Dalai Lama over the Tibetans and contribute towards their assimilation into the Chinese mainstream.

Theatre Development

Roads : Three major highways connect TAR to Mainland China. Of these, the Gormo-Lhasa (also known as Qinghai-Tibet) highway is the most important and termed as 'Tibet's Lifeline'. The other two highways connect East and West Tibet to Yunan Province and Xinjiang Autonomous Region of China respectively. These highways and other roads are being continuously improved as part of Tibet's development and assimilation. They have an extensive road network in the border areas. Under a five-year plan programme (2000-2005) five national and 14 regional highways are to be upgraded and some new ones are to be built at a cost of 10 million Yuans. Work has already started on the first highway on Tibetan plateau from Xining to Tsongkhov Kar forming part of Xining-Lanzhov Expressway.

Railways : Although Tibet is not yet connected to Mainland China by rail, China proposes to link Gormo (Golmud) to Lhasa through a 1118 km rail link by the year 2007. This $ 2.5 bn project involves over 960 km of railway lines being laid at altitudes of over 4000 mtrs. Around seven per cent of the proposed rail link will comprise tunnels and bridges. The longest tunnel at Yangbaijin will be around 1.7 km. When completed, this project will facilitate better integration of Tibet with Mainland China and also lead to Tibet's overall development. This project will considerably enhance PLA's logistics support facilities for additional troops in Tibet.

Airfields : There are three operational airfields in Tibet at Shigatse (Xigaze), Gonghar (Gongga) and Pangta (Bangda). Another airfield is located at Damshung (Donshoon), north of Lhasa, which is reportedly not operational. Many of these airfields are being extended and improved to bigger civil and military aircrafts. In addition to these, there are 15 more airfields around Tibet. Available details are tabulated below *(Source : Tibet 2000, p. 107)*:-

Name	Location	Altitude (ft.)	Type
Within TAR			
Damshung	162 km north of Lhasa	13,123	–
Gongkar (Gonggar)	97 km southeast of Lhasa	11,709	Civil and Military
Pangta (Bangda)	East of Lhasa	14,107	Civil and Military
Shigatse (Xigaze)	40 km east of Shigatse	12,037	Military
Lhasa	10 km north of Lhasa	14,091	Military
Around TAR			
Chabcha	25 km south of Lake Kokonor	10,006	Military
Gangtsa	North of Lake Kokonor	11,601	Minor
Gormo (Golmud)	15 km north west of Gormo	9,275	Civil and Military
Gyalthang	184 km from Dechen	9,000	Civil
Jyekundo	North East TAR	12,831	-
Kandze (Kanze)	NW Sichuan, Kanze Prefecture	11,070	-
Siling (Xining)	South of Siling City	7,211	Civil and Military

A domestic air service within Tibet is also planned and necessary infrastructure is being developed for this in Lhasa area. Such developments would progressively increase the air induction capability into Tibet.

Oil Pipeline : During the 1970s, the Chinese laid an oil pipeline from Gormo to Lhasa. This pipeline is about 1100 km long and meets some of Tibet's requirements of petroleum, oil and lubricants. It is the highest oil pipeline in the world, with a designed capacity of 500,000 tonnes annually but is operating at 25 percent of its capacity.

Military Setup

The TAR is militarily controlled by the Tibet Military District (TMD), which is further sub-divided into six Military Sub Districts (MSDs). However, while Ngari (Ali or Ari) prefecture is administratively under TAR, militarily it is controlled by South Xinjiang MD. There are two types of forces in Tibet. The Border Defence units are permanently deployed on Indo-Tibetan border and are loosely termed as 'Local Forces'. The other troops under the military region or their Group Armies are termed 'Regular Forces'. The approximate strengths are given below *(Source : Study on Deployment of Chinese Occupational Force in Tibet- by Tashi Chutter, Pages 74-75)*:

(a)	Border Defence troops	- 50,000
(b)	Independent Battalions	- 12,000
(c)	Support/Administrative troops	- 20,000
(d)	Regular troops (Two Independent Brigades and Motorized Regiment)	- 26,000
(e)	Air Force troops	- 15,000
(f)	Miscellaneous	- 14,000
(g)	Total	- 1,37,000

All PLA major units in Tibet are being connected by satellite communication. The first section between Lhasa and Lanzhou was established in early 1999. Tibet plans to connect all cities and prefectures by optical fibre cable by the year 2003.

People's Armed Police (PAP) : The PAP is a para-military force responsible for maintenance of internal security in Tibet. It is somewhat similar to India's Central Reserve Police Force. Its organisational structure is similar to the PLA. The approximate strength of PAP in Tibet is as given below :

(a) Lhasa - 15,000

(b) Seven Military Sub Districts (approximately Brigade strength in each) - 38,000

(c) PAP Support Units - 12,000

(d) Total PAP in Tibet - 65,000

The heavy deployment of security forces in Tibet is necessitated by internal unrest and the need for border management. There are periodic reports that the Chinese have stationed missiles in Tibet. While these reports mostly emanate from motivated Western/Tibetan circles, the fact remains that Chinese do have missile sites in Qinghai province and a known nuclear test site in Lop Nor desert.

Integration and Socio Economic Development

The Chinese have been continuously endeavouring to holistically integrate Tibet with the Mainland, while simultaneously degrading Dalai Lama's influence. The year 2001 was celebrated as the 50th anniversary of 'Tibet's liberation'. Besides developing small scale industries, the Chinese authorities have also endeavoured to provide employment to Tibetan youth to prevent their becoming restive. In Tibet the policies are evolved by Tibet Work Forums. The first, second and third forums were held in 1980, 1984 and 1994 respectively. The first and second Work Forums were held during the period of Liberalisation following the Cultural Revolution. The Party Moderates influenced these. The policies set out by the first two Forums were criticised in the Third as being too liberal and favouring Tibetans. It advocated 'fast track economic development and opening up', resulting in migration of Chinese workers/ entrepreneurs into Tibet. It also officially marked the end of an era of liberalisation in Tibet. The Fourth Tibet Work Forum was held in Beijing from 25 to 27 June 2001. It endorsed the general policies laid down by the Third Forum. Development and 'strengthening the administration of religious affairs' to promote Marxist outlook were earmarked as the main priorities.

China's policy on Tibet hinges on "Economic Development and Social Stability" to ensure greater assimilation of Tibet into a 'Unified Chinese State'. A continued focus on 'Party building' was stressed for Communist Party networking in Tibet. The phrase " Leap Over Model of Development" was coined for economic development. This term denotes acceleration in Tibet's economic development. Prior to 1952, Tibet had no modern industry except for a few handicraft units. It now has more than 300 moderate sized modern industrial units. Tourism is being given a big boost. TAR has plans to receive one million domestic and foreign tourists from 2001 to 2005. Chinese Premier Zhu Rongji had announced that the State would invest in 117 construction projects amounting to investment of $ 3.8 billion and other Chinese Provinces would invest in 70 projects totaling $ 1.3 billion.

The Chinese view Tibetan culture, religion and Dalai Lama's influence as a threat to China's stability and detrimental to the policies followed by the Chinese Communist Party. By depicting it as underdeveloped, Beijing has tried to legitimise the Central Government's control over Tibet, despite its autonomous status. Holistic development of Tibet is aimed at weaning Tibetan masses away from the influence of the Dalai Lama. This has been achieved to an extent due to a changed new generation inTibet. The stress on infrastructure development (roads, railway lines and increased air traffic) would better integrate Tibet and increase logistics capability. The extensive development in TAR would provide a round the year stocking capability by rail, road and air.

As China opens up and becomes better integrated with the global system, it may, it is hoped, become less coercive in its Tibet policies. Further, sustained development of Tibet could lead to its opening up to foreign tourism. While the Tibetans may not fully accept Chinese over-lordship, they may progressively become less critical of it. Tibet and the presence of its inspiration, the Dharmshala-based Dalai Lama in India, is likely to remain a thorn in India-China relations, more so if the USA and the West continue to exploit him.

China's wide ranging development activities in Tibet will improve its logistics capability and facilitate speedier induction of additional forces, both from Mainland China into Tibet and move within Tibet from one sector to another. Chinese will be able to concentrate strike formations in comparatively shorter timeframe with much shorter warning period to us. This will provide the PLAAF the ability to strike/engage targets in India on a broad front and in depth. Many of the infrastructural development projects are set to complete by 2006. It is, therefore, fair to presume that by 2010 or so, China's build up capabilities from Mainland to Tibet will be tripled and its induction timings generally would be reduced to half than what they are at present. This will also enable the PLA Air Force to deploy more sophisticated aircraft like SU-27s and SU-30s in Tibet, should the need arise. Under normal conditions also, it will enable China to manage the internal unrest and border affairs more effectively. The importance attached to development of Tibet amply underlines China's sensitivities related to it. In view of the fact that China is in illegal occupation of almost 38,000 sq km territory in Aksai Chin and its maps continue to show Arunachal Pradesh as its territory, there is an urgent need that matching construction of roads and development of these theatres (Ladakh and Arunachal Pradesh) should be carried out so that we are not found wanting.

Chapter 13

China as a Strategic Challenge

Any study concerning China, particularly the one related to its armed forces, invariably raises a major question, that is, whether China is a friend or foe to India? Whether China is a collaborator or a competitor? What sort of threat China poses to India's national security and what response India should generate to meet those threats? Considering the complexity of the Sino-Indian relations in the changing global environment, it is not easy to answer these questions. Nonetheless considering the importance of the issues involved, the trends in Sino-Indian relations in years ahead need deep analysis. India and China are the two earliest civilisations, which in mid 20th Century freed themselves of the yoke of colonialism and imperialism. The two most populous nations attained freedom through different routes and chose disparate paths towards their development. While Indian leaders tried to re-build independent India on the platform of idealism, China adopted a very dynamic, realistic approach to progress. It needs no elaboration that India paid a heavy price for its lack of realism. Historically the relations between India and China have been cordial which were traditionally confined to the exchange of scholars, spread of Buddhism and such other activities which have very little relevance to the present day geo-strategic relations. There has been a long period of isolation while the real contact between India and China is generally restricted to last two centuries or so.

During the last fifty years or after the period when both the countries became independent, India-China relations have not been that warm. The relations generally got the flavour of blow-hot blow-cold from time to time. Prior to 1950s China referred to India as a 'running dog of imperialism'. In early 50s, relations between the two countries deteriorated due to Chinese 'liberation' of Tibet. It subsequently improved to an all level high pitch of 'Hindi-Chinni bhai-bhai' when Chou En-lai made a landmark visit to India culminating in both the Asian giants agreeing to govern their relations on the basis of 'Panchsheel' that is the Five Principles of peaceful co-existence. The post 1962 India-China border conflict led to a freeze in bilateral relations. The clash at Nathula in 1967 hardened the stand on the border problem from both sides but ensured status quo. Although China did not accept merger of Sikkim (1975) with India, there was gradual improvement in Sino-Indian relations in the post-Mao era. However, this did not prevent the Chinese from staging a show of force through the Wangdung incident in India's North-East in June 1986. The overall improvement in Sino-Indian relations came through only with the visit of Rajiv Gandhi in December 1988. This pioneering visit led to the setting up of a Joint Working Group (JWG), which has held 13 meetings till August 2001. The concrete achievements of this august body has been the Peace and Tranquility Accord (1993) and its complementary Confidence Building Measures (CBMs) Agreement (1996). The publicity accorded to China's role in being the rationale for the nuclear explosions in May 1998 led to a temporary downswing in India-China relations, which was apparently checked with the visit of the External Affairs Minister Jaswant Singh to China in June 1999. The subsequent visits by the Indian President to Beijing (May 2000) and Chairman of China's National People's Congress Li Peng to India (January 2001) have lent further stability to bilateral relations. However, major perceptional differences still remain although India is now dealing with China at a level which is significantly different from the 1950s-80s era.

Changing Global-Regional Scenario

There have always been three broad dimensions to Sino-Indian relations i.e. global environment, regional configuration and the bilateral dimension of the mutual relationship.

With the end of Cold War and the demise of erstwhile USSR, the super power rivalry has come to an end. Europe is moving towards a new integration wherein Central Europe is arising again as an important centre of economic and political activity. The world is moving away from confrontation to cooperation. However religious fundamentalism and international terrorism are new threats with global reach which pose a greater challenge to 'haves' than 'have nots'. Whereas Europe may be witnessing an era of peace and stability, the world's focus is shifting towards Asia—West Asia, Taiwan, Kashmir and Afghanistan, which will continue to keep this region in turmoil. The internal situation in Indonesia and the developments in South China Sea involving some of the ASEAN countries and China, if not controlled in time, has the potential to become global flash points. With the recent happenings in Afghanistan, the situation in Pakistan and the Central Asian region is far from satisfactory. The regional environment may not be as benign as the global scene. To have a war in our neighbourhood implies long term new challenges to India's security. China advocates the transition in the world balance of power from the present 'unipolarity' towards multipolarity and views itself, the USA, Russia, European Union and Japan in having the potential to merit status of a 'pole'. China does not consider India as an equal in political, economic or military terms. Chinese economic capabilities are much larger than India and better integrated with Asian economy.

Mutual Perceptions

India and China have viewed themselves as dominant civilisation centres. China's cultural heritage was acknowledged by its bordering states of Mongolia, Korea, Japan and Vietnam. Indian heritage was held in high esteem by South East Asian states of Myanmar, Thailand, Cambodia and Indonesia. China was always a unified state that protected its core (Beijing, Nanjing) by controlling the periphery. Its security strategies had a military orientation. By contract, India was a civilisational construct and not a unified nation-state, historically divided into small kingdoms in conflict with each other. Effective military control was non-existent. Indian and Chinese views of one another are a study in contrasts. Indian perceptions of China are broadly divided into three schools: Centrists, Hawks and Doves. The Centrists do not consider China as an immediate military threat, but are uncertain about its long-term intentions. Hawkish China watchers consider India and China as natural rivals. The Doves view China in a perpetual struggle with the US hegemony and may not be malefic towards India unless provoked.

India receives far less attention in China than China does in India. China's diplomatic engagement of India is a part of its omni-directional Asian policy, wherein India is not perceived as a serious challenge. In consonance with China's Middle Kingdom syndrome, China has always sought to compare itself militarily with the USA and economically with Japan. It has historically tried to deal with India in a smaller regional framework of either Asia or more often in the context of South Asian subcontinent, as an important South Asian country. During the last 50 years or so China's policy towards

India aimed at the following objectives :

(a) Keep India embroiled in South Asian region and its arms-conventional and nuclear–aid to Pakistan provides a low-cost policy option to achieve its long term strategic aim.

(b) To prop up India's smaller neighbours against India's allegedly 'big brotherly' attitude towards these countries while making deep geo-strategic, socio-economic inroads into India's neighbourhood ensuring India's strategic containment.

(c) Continue the rapprochement process with India on all fronts forcing an extremely slow process of CBMs and resolution of the border problem.

Co-operational Aspects

The commonalties and broad similarities between the two nations, which form the basis for co-operation, are linked with their being ancient civilisations, which have made indelible contribution towards mankind and the profound traditional friendship as existed between them from times immemorial. Both the countries have jointly evolved Panchsheel which has been the guiding philosophy for inter-state relations amongst the third world countries. Both the countries hold identical views on many major issues concerning peace and development in the world. Both the countries want to establish a new economic world order. Both India and China need a peaceful internal and externa' environment so that there is a significant improvement in the quality of life of more than two billion people living in these two countries. Thus, both the countries are facing the common task of developing the economy, eliminating poverty and improving the well being of their people. They share the views on many issues such as human rights, environmental protection, combating drug trafficking and population control. The relations between India and China were in a state of dormancy for quite sometime after the 1962 conflict. Since 1988 there had been frequent exchanges at high levels between India and China. President K.R. Narayanan's visit to China and visit of Chinese President Jiang Zemin to India are events of great significance. There appears to be a strong desire on both sides to put the past aside and build multi-faceted bonds of co-operation in order to take advantage of the global progress in economic, technological and social fields. China has often stated that India-China relationship is an important component of its foreign policy of good neighbourliness. As the two most populous countries in the world, India and China have important influence not only in Asia but in the world. Indian and Chinese leaders have agreed to establish a cooperative and constructive partnership on the basis of the five principles of peaceful co-existence.

China is the home to about 16 million Muslims of various ethno-national minorities, located in strategically important parts of China. Like India, China is also facing Islamic militancy which is fuelling separatism in the Xinjiang province inhabited by Uighur Muslims. For the stability and peace in Asia, India and China have a lot in common, to join hands in fighting the three evil forces i.e. international terrorism, religious extremism and ethno-nationalist secessionism. As multi-religious, multi-ethnic and multi-lingual states, both have a mutuality of interest in countering such challenges. China is the most populous country in the world and India comes second. Both the countries have two major challenges i.e. to check the growth of population as well as to give a decent standard of living to their teeming millions. To check population growth is relevant to economic development essential for social stability and ecological balance in Asia. Thus, with proper understanding, India and China can be partners in economic prosperity and world peace so as to carve out a progressive new world economic order which will ensure the resurgence of Asia in years to come.

After Kargil operations, China clearly described Kashmir as a bilateral dispute between India and Pakistan and urged both the parties to resolve it under the framework of Simla Agreement. Also it was the persuasion by China that resulted in Pakistan withdrawing its resolution on Kashmir at the UN Human Rights Commission's annual session at Geneva in March 1996. Similarly in January 1995, when the Western powers had withdrawn the promise of supplying missile fuel for India's Tarapore Nuclear Plant, it was China that came forward and supplied one installment of enriched Uranium. At global and regional levels both the countries have evolved a working relationship as long as it remains within China's overall policy framework towards India, that is, while China is a regional power with global reach and India is a South Asian power with regional reach.

India-China Trade

Sino-Indian trade has never been significant due to geographic barriers, which are not easy to surmount. The pattern of the volume of bilateral trade during the last two decades or so is as follows :

Year	Volume of Trade (US $)
1977	25 million
1984	62 million
1992	340 million
2000	2 billion
2002	3 billion

* Source: Express India, Indian Express, PTI Report in http:// www. expressindia. com/ fullstory.php?newsid=6500

During the period 1979-1999, while China's international trade has gone up by 14 times, Sino-Indian trade has increased by 16 times during the same period. Due to high-level political interaction during the last decade or so, Sino-Indian trade has expanded rapidly with the annual trade volume at present amounting to over $ 2 billion coupled with the development of border trade. Cooperation in Science and Technology and Information Technology as well as exchange of economic and cultural delegations have considerably expanded in the recent past. Both China and India have already set up over 50 joint ventures in various fields. However, this represents a small portion of the overall trade volume of the two countries. Undoubtedly, there is scope of further expansion. With concerted efforts from both sides, the volume of trade should cross $ 5 billion within next two years or so. India is rich in iron ore resources, the bulk of which is high grade and has low exploitation cost. China has a huge demand for iron ore. There are excellent prospects for Sino-Indian joint ventures in iron and steel. Another major area of cooperation is science and technology, particularly in the field of aerospace. India is more advanced in developing computer software whereas China has excelled in hardware. This is an important area where the two countries can upgrade their cooperation and exchanges to their mutual advantage. Agriculture and agro-industry is another area in which China has achieved considerable success in seeds and other agricultural sciences. India could take benefit of the long strides in agriculture which China has already achieved. The enhancement in trade needs to be viewed as a CBM resulting in greater interaction and mutual confidence not only in the border communities but the top business houses from both sides. Even after Pokhran II, scholars from China and India have continued to advocate sub-regional cooperation between China's South-Western, India's North-

Eastern region and Myanmar by creating sufficient infrastructure for the development of the areas which are economically backward and neglected through ages. In the long run, expansion of mutual trade and investment will help in creating mutual commitment ensuring peace and security between India and China. Unfortunately at present even the improved political relations lack enough material basis and the improvement of security relations lack enough economic impetus. The continued stalemate and perpetually existing mutual suspicion will require bold political initiatives from both sides, if the trade relations are really to take-off.

CONFRONTATIONAL ASPECTS

Regional Rivalry

China's aspiration to attain global power status is well known. As a permanent member of UN Security Council and as the fastest developing economy in the world, China has largely been successful in attaining regional pre-eminence particularly with the demise of erstwhile USSR. In its quest for Asian leadership and global influence, China finds that India is the only country that could come in the way of its aspirations. Thus many Chinese thinkers consider India a strategic challenge. This challenge necessarily need not translate into a military threat, at least in the near future. China has followed a three-tier defensive strategy vis-à-vis India. It maintains a relatively small force level on India-China border, well within its logistics constraints. China has skillfully committed its nuclear forces, targeting Indian cities as a strategic insurance. This compensates China's conventional vulnerability vis-à-vis India. The third component of China's India strategy is through proliferating and bestowing nuclear capability on Pakistan to keep India's focus on Pak. Dr Tellis assesses that India-China security interdependence is likely to manifest through the struggle for security, accumulation of wealth and recognition of status. India named China as the *raision d'etre* for its Pokhran II nuclear tests in May 1998. China took umbrage and the bilateral relations suffered till damage control was initiated by India. Though the issue is now on back burner, China continues to be adamant about India signing NPT and CTBT.

Both India and China are in favour of a new world order, that is, a shift from uni-polar American dominated world to a multi-polar polity in which both the countries should find prominent role in global affairs in general and Asia in particular. However, their views significantly differ on what the new world order will be. India's world view is polycentric while China is looking towards a multi-polar world order. Next to the USA, China aims to become another major 'pole' as the erstwhile USSR used to be during the cold war period. The multi-polar concept implies regional hegemony whereas India's polycentricism disapproves of all hegemonies, regional or global. Thus there is a significant difference in their perception of the emerging post cold war international balance of power.

There are clear manifestations of lack of understanding and enduring stability in Sino-Indian relations. China's strategic aims are directed at the achievement of its territorial and economic security by limiting that of its potential rivals. China is not likely to accept the growth of any perceived regional rival, particularly along its south and south-western borders. China has over the years been steadily increasing its influence in South Asia by befriending India's neighbours. Thus China is apparently aiming to control both the eastern and western flanks of India through Myanmar and Pakistan. Though India is not China's main adversary, the importance it has given to its strategic environment indicates her desire to establish her dominance in the region. The growing US-

Indian relationship could be another irritant towards the stability of India-China relations. Despite recent improvements in India-China relations and border CBMs, India remains suspicious of China and its growing regional and global influence. This mutual suspicion and lack of basic trust is likely to persist for a long time to come. China is our definite competitor in economic field. The sheer size, proximity and the potential of these two nations are bound to manifest in a clash of interests. Economic compulsions may preclude any major conflict in short term. In foreseeable future there is no possibility of China becoming India's ally. While there might be convergence of interests in global affairs, in the Asian region China and India are potential rivals.

Border Dispute

The boundary question remains the most central and toughest of all issues and the slow progress here will continue to delay the overall progress in India-China relations. The 3917 km long India-China border, was demarcated along the British drawn McMahon Line in 1914. The treaty was not ratified by the Chinese at that time. After both countries became independent, while India thought that the border is already demarcated, Chinese perception was that the boundary was the legacy of the colonial era and needs to be renegotiated. The Sino-Indian border is divided into three sectors – Western, Middle and Eastern. While the 400 mile long Middle Sector covering Himachal and Uttar Pradesh is relatively non contentious, the problem areas are in the Western and Eastern Sectors. The 1100 mile-long Western Sector that contains 15000 sq.km of Aksai Chin also contains about 2700 sq. miles of area that Pakistan ceded to China from Pak Occupied Kashmir in March 1963. The Eastern Sector is 140 mile-long boundary along Sikkim and 710 miles-along Arunachal Pradesh. Despite the fact that Sikkim acceded to India in 1974, Chinese maps continue to portray it as an independent country. During the 1962 War, the Chinese seized an area roughly the size of Switzerland. In the Eastern Sector, the Chinese over-ran the state of Arunachal Pradesh and threatened the plains of Assam. After the cease-fire, the Chinese went back to the Line of Actual Control (LAC) which generally approximates to the McMahon Line. The disputed area, however, is 33500 kms in the Western Sector, 2000 sq.km in Central Sector and 90000 sq.km in the Eastern Sector. As many as 13 Joint Working Group (JWG) meetings on the border issue have taken place since 1989, but hardly any progress has been achieved. After the two accords in 1993 and 1996, maps of the Middle Sector were exchanged between the Chinese and Indian authorities in November 2000 and discussed during July 2001. China continues to improve its infrastructure in Tibet and the areas adjoining pockets on the border where the LAC is perceived differently. Perhaps it is to present de-facto control over these areas and seek bargaining chips in the unending border negotiations. While India continues to seek some progress as a result of these JWG meetings and high level talks, China it appears is encoded in its belief of growing world power status, and is content with keeping the schedule and pace to suit its long term "designs".

It is interesting to note that China inherited territoral dispute with many countries including Russia, Myanmar and Vietnam. Other than India, most of the territorial disputes have been settled. For resolving these border problems China has adopted the strategy of Calculative Security approach to secure its basic interests. In case the territorial dispute is trivial to China's larger interests, China has sought to solve it amicably, for example, resolution on Sino-Russian and Sino-Kazakhstan and Kyrgystan border disputes. In case the dispute is significant but cannot be peacefully resolved to China's advantage at present juncture, it is indefinitely postponed through long-term negotiations. Two instances of this are China's dispute with India and Japan. This strategy enables China to avoid conceding its claims, while maintaining the placid peripheral environment it needs. This also projects China's image as a conciliatory state and prevents use of military force from the position of weakness. It also delays resolution of the dispute till the balance of power changes in China's favour. Whether a strong China will continue the policy of indefinite postponement remains a moot issue. China has preserved status-quo without renouncing its claims. However, it has not hesitated to use force, as in the case of its war against India, Vietnam and even Russia—whenever China perceived that the status-quo is likely to disturb the existing arrangement to its disadvantage. Many Indian strategists legitimately think that China is dragging the border dispute to some future date when it will be able to settle it to its advantage from a position of strength.

The dismemberment of power blocks and trends towards globalisation, there is increasing awareness both in China and India that both the countries must settle their problems amicably, end the era of uneasy peace and move ahead towards the betterment of the people. From India's point of view the core 'issue' is the border dispute. Therefore, any lasting peace between India and China will depend on the resolution of border dispute. Both sides clearly understand that the dispute can only be resolved through mutual accommodation. However, there is no indication of what constitutes a reasonable compromise. The perception of both sides are widely apart. While India seems to think that it is for China to show greater accommodation, the Chinese believe that they have lost thousands of kilometres of territory, so the concessions must come from the Indian side. It is generally believed in India that the Eastern Sector is virtually settled and for the settlement in Western Sector the Chinese should make some concessions. Beijing has taken a very different view. The statements and writings of many Chinese military thinkers reveal that China considers Eastern Sector, specially Arunachal Pradesh, as the key issue. Some high ranking Chinese officials while interacting with Indian journalists clearly stated that in the border dispute India has not lost very much and it was China which has lost thousands of kilometres of its territory. According to them, the concessions, if any, are needed from the Indian side. Such thinking shows that the border dispute is likely to linger on for a long time to come. With various CBMs in place, the unsettled border necessarily does not imply the tense or militarily active border. For next two or three decades, China needs peaceful surroundings for its economic growth and development. It is, therefore, likely to continue with its Calculative Security Strategy till it acquires comprehensive national strength. China's present key objectives are economic development, domestic order, peripheral security and geo-political recognition. As its capabilities increase this could be enlarged to exert more control over its surroundings. Any shift of its Calculative Security Strategy may also be dictated by the level of dependence on international economy and external environment related to foreign investment, energy resources and the role which the United States of America will play in Asian affairs.

Some important developments are taking place in the Indian Ocean which need to be taken note of. In a press conference in November 2000 (*Asian Age* November 2000) Indian Navy Chief Admiral Sushil Kumar warned of increasing conflicts, tensions and crisis in the Indian Ocean. He stated that Indian Ocean will become the major outlet of fossil fuel reserve triggering the tussle for supremacy which could create tension in the region. China lately has been exerting greater influence towards the Indian Ocean than ever in the past. One of the remarkable and potentially dangerous development in the post-cold war era is the rise of concept of sea as national territory. Lt. Col. Chu Yu Chen in his book *A New Scramble for Soft Frontiers* wrote that China's area for survival is shrinking. He adds that "we have to reclaim sovereignty and sovereign interests in the oceans, territorial sea, continental shelf and exclusive economic zones", an area which it believes was lost to the erstwhile colonial powers like Japan, Russia, France and Britain in the last two centuries. The Chinese people have been prodded to cultivate and nurture this new concept. People's attention is directed towards the unpleasant fact that in recent history it is China's maritime interests that have been encroached upon, resulting in great loss to China. Chinese strategic thinkers often stress that "territorial claims laid by foreign countries over China's maritime territory amount upto one million square kilometers, ten times the size of China's disputed land borders. By now, over thirty Chinese islands with surrounding waters are still in the hands of foreigners. Losses in maritime interests are really to be grieved and they call for our serious consideration." The Chinese strategists now speak of the need for survival space and for strategic frontiers that extend horizontally into the Indian Ocean, the South China and East China Seas and vertically into space. In keeping with such a concept of sea as national territory, China's naval military doctrine has shifted from the coastal defence of the mainland to active defence of maritime, economic, resource and strategic interests. The security of Indian Ocean sea lanes for the smooth flow of Gulf oil is of great importance for China. The International Energy Agency in Paris estimated that Chinese oil imports, mainly from West Asia would rise to more than 8 million barrels a day by 2020, increasing China's net imports dependence to around 76 per cent from about 25 per cent now. China also feels that the US forward military presence in Asia in the next one or two decades will decline. Such a situation will pave the way for a more significant role for the Chinese Navy in the Indian Ocean. The mid-term goals of PLA modernisation like Rapid Deployment Force, air-to-air refuelling and improved logistics support system are being targeted to achieve such limited power projection capabilities to support its emerging concept of sea as national territory.

Tibet and Dalai Lama

India is a home to more than 100,000 exiled Tibetans, while the Dalai Lama's Government-in-exile functions from Dharamshala. The Indian Government recognises Tibet as an autonomous region of China and does not permit political activities inimical to China on Indian soil. The Chinese perceive exiled Tibetan activity as secessionist, which poses a direct threat to the stability of Tibet and security of South East China. The recent Panchen Lama controversy is a clear indication that China is determined to undermine the ethnic, cultural and religious identity of Tibet. The Chinese are preparing for a post Dalai Lama Tibet. Occasional support by Indian pro-Tibet lobby to remission of status of Tibet is seen by China as a measure to exert pressure and influence China. Hence, India's role in Tibet is viewed by China with suspicion and this aspect will continue to remain an irritant in Sino-Indian relations for a long time to come. It is reasonable to assume that India may see an increase in the Chinese intelligence activities on its soil.

Pak Factor

India's suspicion of China's relations with Pakistan, over last 50 years, are more than justified. In fact propping up of

Pakistan against India is a 'low cost option' for India's containment which serves China's long term geo-strategic interests. China wants India to remain grounded in South Asia. India's Pak-centric obsession gives China a greater freedom of action in Asia and helps her in achieving her strategic aims better without any competition from India. China believes it is a big power and in foreign relations particularly in Asia, it behaves as such. A big power does not abandon its allies or those who are dependent on it. Therefore, China will not abandon Pakistan but as has been evident from its policy pronouncements during the 1999 Kargil War, it will pursue more balanced relations on the South Asian sub-continent. The Indo-Pak conflict has provided China with a lesson on what constitutes responsible behaviour in international affairs. What with Pak's role in the export of Islamic fundamentalism and the trouble which China is anticipating from its Muslim minority in its sensitive western region having borders with PoK and Afghanistan, it is probable that Beijing's support to Pakistan, outside India-bashing domain, has to be somewhat moderated by the fear of the rise of similar fundamentalism in the other regions of China. In a recent statement on Sino-Pak relations, Chinese Foreign Minister stated that Pakistan is not China's ally but China's neighbour and its time tested 'all weather' friendship will get further strengthened in the years to come. China will continue to play the Pak factor as long as it serves its strategic interests.

Security Behaviour

Historically China has been a continental country and much of its security concern was related to Inner Asia. Like most of the ancient civilisations, China's security strategy has focused on consolidation of control over its far flung and vulnerable periphery. China normally relied upon a strong

Beijing based central authority with frequent and intense application of coercive force coupled with aggressive diplomacy, traditional culture and a progressive, decentralised economic system. When in a position of strength, China exercised strict control over its strategic periphery through establishment of suzerainty-based relations backed by superior military force. However, when the central authority in China was weak, it opted for non-coercive strategies like appeasement, alliances, cultural and personality-oriented interactions and heavy reliance on static defences. An analysis of Chinese strategic thought suggests that China has usually preferred to eliminate its opponents through offensive force rather than static defences or accommodation, when it clearly enjoyed a superior military position. This pragmatic policy has not been abandoned. Some Chinese historians have argued that unlike their Western counterparts, Chinese rulers were historically non-imperialistic and did not conquer surrounding 'barbarian' countries even when they could do so without much difficulty. They argue that the only real instances of territorial expansion took place when non-Chinese Mongols were in power in China. After its independence China has the unique distinction of fighting a war with both the super powers and major neighbours in Asia–India and Vietnam. It did not hesitate to use force against its South East Asian neighbours, with whom she has generally had cordial relations, over its dispute in South China Sea. In essence, China's foreign relations have displayed wide fluctuations particularly with regard to Asian countries. China's relations with India transformed from one of brotherliness to a War in 1962; in the mid 1960s its relations with Indonesia abruptly shifted from partnership to extreme hostility and in late 1970s Sino-Vietnamese relations quickly deteriorated from 'boundless affection and assistance' to hot, and then cold war. China's security behaviour seems to be driven by unilateralism based on its perception of being the most important power in the Asian region. It has insisted on all its territorial disputes being resolved directly on bilateral basis without any interference by any third country or international agency. In handling its territorial and maritime disputes, which are numerous, China has taken an aggressive head-on approach. It clearly emerges from recent Chinese internal military writings that war is still considered preferable to any appearance of surrendering sovereign, territorial claims. The disputed Paracel and Spratly Island groups is a dangerous flashpoint in the Asia-Pacific region. No less than six states – Brunei, China, Malaysia, the Philippines, Taiwan and Vietnam – have competing jurisdictional claims over the potentially oil-rich Spratly Islands. This Group of Islands also straddle sea lanes vital to Asia-Pacific states, adding geo-strategic and geo-economic dimensions to the simmering conflicts. China, Taiwan and Japan are also locked in dispute over the Diaoyu (Senkaku in Japanese) Islands farther north in the East China Sea. To possess the Diaoyu/Senkaku Islands, which comprise five islands some 166 km north east of Taiwan, is to have legal jurisdiction over about 21,645 square kilometres of the continental shelf. This is believed to be one of the last unexplored hydrocarbon resource areas in the world. In 1994, Indonesia discovered that according to China's claims, its Natuna gas field, well within its 200-mile exclusive economic zone, lies within Chinese waters. In 1995, Filipinos woke up to find a Chinese flag fluttering on Mischief Reef just 200 kilometres from Palawan Island, despite the fact that the Reef is well within Filipinos 200-mile exclusive economic zone duly recognised by the United Nations Convention on the Law of the Sea, while the nearest inhabited Chinese territory is Hainar Island more than 1000 miles away to the north. As per their ethos, the Chinese have practiced 'creeping assertiveness' in thickening their presence in the disputed areas irrespective of the fact whether it is at sea or on land. As China grows in military and economic strength, she will behave boldly, more inclined to force its will upon others –coercive rather than consultative. The Chinese security behaviour clearly indicates that at an appropriate time a stronger China may re-define its strategic interests. This is an uncomfortable thought for India with a lingering border dispute with China without any plausible solution in sight.

Clash of Interests

In the long term, China will continue to perceive India as its potential strategic rival in Asia. India's nuclear capability would be factored into her power projection calculations. For its continuous growth, China's developing economy would require greater share of South East Asian and Asian markets. This is a region where India traditionally had and will continue to enhance its goodwill and socio-economic influence. The clash of primary economic strategic interests between India and China may create a conflictual situation. Similarly, China's increasing interests in the Indian Ocean coupled with its desire for a blue water navy are dangerous indicators for developments ahead particularly for India and Japan. To become a global power, China will try to maintain its ascendancy in the Asian region. If at any stage it feels that its economic or military capabilities are being challenged, it may choose to display its power superiority across disputed borders through a swift military action 'to cut India to its size'. China's foreign policy will be greatly influenced by its internal developments. Any destabilising internal unrest, particularly in Tibet, may force China to inflict a blow on India to demonstrate its national will and overall superiority. China's history of 'teaching a lesson' bears testimony to its unilateral security behaviour. Such an action would confirm to the world the 'arrival' of China in the super power league. How aggressive China may become depends on many imponderables which may be difficult to predict at this stage but it could be stated with reasonable certainty that on Asian horizon India and China will co-exist as rivals rather than allies. It necessarily does not mean an open confrontation but numerous unpredictable situations may arise having potential to develop into a conflict which India can ill-afford to ignore.

In the short term, China needs peace on its borders to hasten its economic development. To that extent its relations with India are likely to be cordial without much warmth, outwardly accommodative and flexible. China's views regarding India will largely be influenced by India's posture on Tibet, the border dispute and the emerging Indo-US relationship. China will continue with its covert and overt policy of India's strategic containment through its neighbours thus impinging on India's sphere of influence without blatantly being hostile, aggressive or uncompromising. Militarily it will force India to keep its armed forces stretched on both its fronts, that is Pakistan and China, while combating terrorism and insurgencies in Jammu and Kashmir and the North-Eastern region. China retains the option of fueling anti-India propaganda in projecting India as a regional hegemony with collusive support from India's neighbours. It can destabilise India by increasing its clandestine support to insurgents in North East and resort to supply of weapons and ammunition through gun runners operating from Thailand, Myanmar, Bangladesh, Nepal and Pakistan. Without crossing the nuclear threshold, it can activate the Indo-Tibet border to India's disadvantage.

China's Military Objectives

In order to ensure its undisputed leadership of Asia, China will continue in its efforts to destabilise and weaken India, while maintaining a façade of improving relations and behaving as a 'responsible' peace-loving nation. She would like to achieve her aim by non-military means including extensive use of coercive diplomacy, trade and economic blackmail. However, there could be situations in which China could resort to armed action of varying degrees, depending on her overall aims and intentions. In specific terms such a threat may manifest itself in any of the following forms :

(a) Occupy unheld or lightly held areas close to the Line of Actual Control.

(b) Capture Indian territory upto her claim line including well known population centres like Tawang, Walong, Tuting, Bomdila upto the foothills and thus threaten the oil-rich plains of Assam.

(c) Threaten important strategic areas even beyond her claim line to include Chushul in the Western sector, Badrinath in the Central sector, Sikkim and Siliguri corridor in the Eastern sector thus severing the North Eastern part from the rest of India.

(d) In the event of a general war against India, China may develop its offensive through Bhutan, Nepal and Myanmar for the capture of North-Eastern region. In such situations she would endeavour to obtain collusive support from Pakistan, and if possible, even from Bangladesh.

(e) With a blue water Navy and greater presence in the Indian Ocean, China will be in a position to :

  (i) Secure Indian Island territories.

  (ii) Pose a threat to Indian sea trade as well as supplies of crude from the Gulf.

  (iii) Exploit the resource rich Indian Ocean region, which would clash with Indian activities in its EEZ or edioining areas.

Most of such actions could be undertaken without crossing the nuclear threshold. The possibility of the Chinese prolonging the war, to test India's sustaining power, cannot be ruled out. There can hardly be any doubt that considering China's over-all force level, in the event of a major offensive against India, she will have no difficulty in mobilising sufficient forces for the attainment of its strategic aims. However, her major constraint could only be lack of logistics infrastructure to support the requisite quantum of troops. During last decade or so, China has taken some important deliberate actions for the improvement of roads, railways, airfields and oil pipeline to enhance its logistic capabilities in Tibet. It is estimated that the Chinese normally hold adequate stockpiles in Tibet for the maintenance of a force level of ten divisions including its Air Force elements for a period of one year. With improved logistics infrastructure, it should be possible for China to induct and maintain a force level of approximately 30 divisions for launching two to three major offensives against India, depending on her overall aims. It needs to be noted that the best approaches into India lie through Northern Myanmar, where the passes are open throughout the year. It is believed that old Ledo Road is being improved. The other route, the Tamu-Imphal axis, is also available. Development of operations through Myanmar would give China an option to tackle Indian defences from the rear, an unexpected move which will cut entire North East from the rest of the country. Depending on what the Chinese wish to achieve, it should not be difficult for them to muster such a force level nor to maintain them through Tibet and/or Myanmar for launching a major offensive against India. While the chances of an armed confrontation between India and China in the short term is remote, the possibility of a conflictual situation in the long term cannot be ruled out. An economically developed China with its modernised armed forces shall not be vulnerable to foreign pressure. It will not be wise to depend on the US nor a revived Russia could be expected to take on China for India's sake. China is a long term strategic threat to India. How the threat will manifest itself in the years ahead cannot be clearly anticipated. It is in India's interest to generate adequate responses – economic, diplomatic and military – to long term threats from China. Whilst a belligerent China may be an unlikely eventuality in the near future, genuine India-China bonhomie too may remain elusive. China needs to be viewed by India neither as a friend nor foe but a gigantic neighbour, and a strategic challenge.

Conclusion

The aspect of China remaining a strategic challenge to India will depend upon international, regional and internal issues. Economic prosperity and political freedom are inseparable. The denial of political pluralism is unsustainable in this information age and globalised economy. China has the maximum number of foreign students in the USA. Therefore to balance the transition from its present dictatorial one party rule to pluralism in a smooth graduated manner is a major challenge for China. The evolving new world order and the US's attitude towards it may determine India-China equations at global level. At regional level, competition over spheres of influence in South Asia may persist. Whilst this thesis may hold good till China acquires requisite 'Comprehensive National Power', the fact that India and China will continue to vie for possibly conflicting spheres of influence in Asia cannot be wished away. An economically and militarily stronger China rubbing shoulders with the USA, may not be content with its status quoist security strategy. The manner in which China then addresses the long-standing India-China boundary dispute is the dilemma for which India must prepare itself well in time. If the existing 'Three Agreements' (Panchsheel of April 1954, Peace and Tranquility Agreement of September 1993, and Confidence Building Measures Agreement of November 1996) are genuinely progressed, Sino-Indian relations could become somewhat less complicated. It would be in the interest of both nations to strive for this ideal.

Chapter 14

India's Response

In the long term China's growing national strength – economic and military – will create security challenges for India, which need to be catered for well in time. The modernisation of the Chinese Armed Forces will tilt the balance of nuclear and conventional power in Asia in favour of China to India's great disadvantage. China has viewed military power as the most important component of the comprehensive national strength which is considered indispensable to regain and retain its global power status and to safeguard its vital interests against any anticipated actual or imaginary threats. Power flows through the barrel of the gun, nationally or internationally, is a time tested dictum of Mao, which has not lost its relevance. Such traditional thoughts are well embedded in Chinese psyche. The situation on India's neighbourhood being what it is, faced with the possibility of a limited war with a 'nuclear overhand', peace can only be achieved by ensuring credible deterrent capabilities – both nuclear as well as conventional. Effective deterrence must aim to raise the costs of military adventurism to levels that would dissuade the aggressor from exercising such options. Militarily, India needs to maintain parity with China at conventional level and an assured second strike capability in nuclear, chemical and biological warfare to ensure its long term security. In our response to threat from China, we will do well to learn some lessons from China in the same way as she is preparing herself to face a long term threat from the United States of America. China is developing comprehensive national strength while keeping the US engaged at all levels despite the latter's security commitment to Taiwan and Japan. Like China we have to progress economically and militarily. In specific terms there are some steps which need to be taken.

Economic Development and Good Governance. It is China's 'economic wonders' which has brought it to centre stage in global affairs. China fully understood that without economic progress no other worthwhile military or social development is possible. Despite being a Communist country, China has managed the market economy and globalisation much better than democratic India. Even after 50 years of independence more than one-third of the Indians live on Rs. 10 per day. Black money forms 40 per cent of our GDP. The Government losses due to tax evasion annually amount to Rs. 2.5 lac crores. Our power transmission losses are 27 per cent as compared to three per cent in China.

Multinationals have poured in more than $ 300 billion into China over the past decade compared to a paltry $ 18 billion into India. India has the potential to attract FDI worth $ 100 billion over the next five years. However due to rampant corruption, lack of infrastructure and bureaucratic red-tapism it is unlikely to happen. For the past two decades, despite the global economic slow down, the Chinese economy has compounded at an economic growth rate of 8 per cent in real terms. In 2000-2001 Chinese economy topped a trillion dollars. Another two decades of a similar growth profile, which is quite plausible, will see the Chinese economy at par with the US economy. Though somewhat exaggerated, Lt. Gen. Hamid Gul, former Pak ISI Chief, has stated that "China is US's new enemy and China is biding time. For the next 20 years, it is focussing on its economy. Twenty years down the line, its economy is going to be worth $ 16 trillion per year. The US economy will be worth $ 8.3

trillion. That's when China will begin to play a new role" (*The Times of India,* 21 December 2001). While the figures may be disputed, the trend is unexceptionable. The large number of prosperous non-resident Chinese all over the world, with a sense of fanatic nationalism, are a source of great strength for mainland China. China attracts 20 times the FDI India does. Despite global economic slow down, Indian economy is also progressing. However, when compared to China, in so far as it relates to over-all economic development, we have already been left far behind. Though somewhat coercive and undemocratic, China has reaped rich dividends by checking its population growth through the one-child policy. India's uncontrolled population growth besides large influx of migrants from Bangladesh, Sri Lanka and Nepal, has been a big drain on its economy. Unless we have double digit economic growth and good governance for at least a decade or more, we will continue to lag behind China in every field. Economic growth and good governance are deeply linked with political stability which may not happen considering the emerging political trends in our country. Growing regionalism, casteism and communalism are major threats to our political stability and consequently economic development. We have to put our economic house in order. To keep the gap between India and China within reasonable limits, our economic growth has to be much more than the present rate, which appears to be a dream with frequent downward revisions of expected growth rate in GDP. In case we are unable to sustain an annual growth rate of atleast Eight per cent or more, then 20 years hence, India and China may be in different economic leagues–developing and developed–with attendant disastrous ramifications. Notwithstanding achievements in information technology, without a dynamic and rapidly growing economy, India will not be taken seriously as a significant Asian, let alone a global power. Without economic power, India can have no security, even with nuclear weapons. An economically weak India shall have no voice in the comity of nations, be it at regional or global level.

India's Attitude Towards China

China is our powerful neighbour. While you can fudge with history, you can't change geography. Because of the humiliation suffered by us in the 1962 Sino-Indian War, we are over-cautious, super sensitive, passive and generally reactive in our policy initiatives towards China. The 1962 psyche and mindset must no longer govern our attitude towards China. In handling China, the deep sense of Indo pessimism should be replaced by Indo realism. Without being excessively aggressive or war-mongering, we must work towards changing the prevalent impression that India is a 'soft state'. We need to create national will and national spirit through which our nation is geared to make all sacrifices to safeguard our core values and national interests. We should evolve a long term policy towards China based on hard geo-strategic realities, backed up by total national consensus. We should maintain business like approach in our dealings with China and harbour no sentiments. Chinese civilisation is essentially scientific in character whereas ours is based on spiritualism and tolerance. As a nation we lack competitive spirit and killer instinct. Therefore, there is serious mismatch to our disadvantage.

Many of our problems with China are historic in nature and will defy any quick-fix solution. We need consistent, pragmatic and patient approach in handling our problems with China. It is generally accepted that there can be no settlement of the boundary question without compromise by both sides. Towards this end, there is a need in India for an all-parties consensus on the likely contours of settlement with China. A possible approach could be to consider de facto control as the starting point, with minor negotiated alterations. Other general principles of settlement could be exchange of territory vital to a country and exchange of settled population would not be undertaken. Acceptance of the principle that neither side should gain from the 1962 War would be an important step. Thus, the two countries would need to abandon purely declaratory approach in favour of structured dialogue, which would lead to practical solutions. Settling the border problem, equitably and early, must be the corner-stone of India's China policy.

Tibet will continue to play an important role in India-China relations. The Indian government has often reiterated its policy that Tibet is an Autonomous Region of China and that anti-China political activity by Tibetans is not permitted on Indian soil. Beyond this, it is not feasible for India to accommodate Chinese wishes on presence of Dalai Lama in India and his worldwide popularity. Presently, China accepts this Indian position. However, it may not be so if bilateral relations deteriorate. Tibet is an important component of China's perceived integrity, unity and security. It provides a major buffer on China's south west flank. It is an unexplored region, rich in minerals and other natural resources. China is trying to alter the demographic pattern in Tibet to its advantage. It is gaining time to consolidate its control, though it is not an easy proposition. The Tibetans cannot get rid of the Chinese and they are too large, dispersed and spirited to be forced to accept the Chinese rule. The struggle will continue unless a mutually acceptable solution is found. The most sensible course obviously is a dialogue between Dalai Lama and the Chinese Government to work out a modus vivendi. Under the circumstances, India does not have many options on Tibet. No basic change in our policy on Tibet is required. However, we must create an environment in which it is possible for us to have a closer look at Tibet. Initially the

Chinese might be hesitant to open up Tibet to us, but with other CBMs being put in place and with consistent efforts we should be able to make some headway. The thrusts that could be considered are as follows :

(a) Establish diplomatic presence in Tibet by setting up a consulate at Lhasa.

(b) Increased trade with Tibet which will ensure greater interaction and dependence of that region on India.

(c) More visits to Tibet by various commercial, religious and military delegations.

(d) Promotion of tourism both through land as well as air routes.

Modernisation of the Indian Armed Forces

The modernisation and upgradation of the defence services involving the military hardware, security apparatus and personnel is a constant and continuous process in most of the developed and developing countries in the world. In the aftermath of the 'Desert Storm' particularly, there has been a technology based global Revolution in Military Affairs (RMA). China has recognised the potential of RMA and has viewed it as a route to improve its relative position against its potential enemy that is the USA in any showdown over Taiwan. However, it does not imply that China will be able to catch up with USA in military capabilities in the foreseeable future. RMA technology is a complicated hi-tech affair. These are neither easy to master nor easy to operate. China will face many barriers, human, financial and technological to master competence to utilise RMA, high grade technologies. For that matter no Asian nation including Japan may be able to achieve this with in the next 10-15 years. Nonetheless a well phased modernisation programme for PLA spread over two decades or so will widen the gap in terms of conventional warfare capabilities to India's great disadvantage unless priority is given to modernisation of India's Armed Forces, and this is treated as a 'national programme'. While we have a fairly strong defensive posture against China, our ability to conduct offensive operations in mountainous rugged terrain through well poised, properly trained and dedicated strike formations is highly restricted. In nuclear and conventional missiles India is almost a generation behind China. Our defence industrial growth has been dismal, considering our shortages in critical arms, ammunition, force multipliers and support systems. The effectiveness of any armed forces rests on the state and vintage of its equipment, which must be upgraded or replaced at regular intervals in a phased manner. Unfortunately it has not happened. The Army needs replacement of old tanks and guns. Besides the replacement of the aircraft carrier, Vikrant, the Navy has a backlog of nearly 25 ships. The Air Force badly requires Advanced Jet Trainers as well as replacement of the Mirages and old MIGs. Our development and upgradation programmes are running much beyond the desired schedule. The LCA and the T-72 tanks upgradation programmes are instances. Our inter-service integration and macro decision making process is far from satisfactory. Even after 50 years of independence we are still debating about the necessity of having a permanent Chief of Defence Staff (CDS) and the integration between the Ministry of Defence (MOD) and Services Headquarters. We took too long to set up a National Security Council (NSC). During the Kargil war with Pakistan, not a single meeting of the fully composed National Security Council (NSC) took place. In real terms India's defence spending in the last decade has gone down despite significant increases both by China and Pakistan. Our defence and foreign policies so far have been Pak-centric. Handling China is a different ball-game altogether. A nation of the size of India must clearly define its security parameters and even make transparent its pattern of future response to the type of mischief which may erupt on its land and maritime boundaries. For the creation of sufficient military strength the following measures need to be implemented :

(a) India has not fully grasped the potential of the current global RMA. The tactical and technical upgradation of the Indian Armed Forces has rather been slow. It is basically an improved version of a largely manpower oriented armed forces of World War II vintage. Our defence industrial base is rather primitive. We are still dependent on imports for our major weapon systems, combat aircrafts and submarines. While we need not match bomb for bomb and ship for ship with China, it is imperative to maintain reasonable parity for our national security and regional status in Asia. Modernisation of armed forces is a long drawn process. The pace of the modernisation and upgradation of Indian Armed Forces need to be accelerated. We must have a permanent CDS for better integration and co-ordination of all the services, nuclear weapons and DRDO. The long awaited integration of MOD with armed forces Headquarters need to be implemented vigorously. At present MOD has all the authority without any accountability.

(b) Induction of new weapons and equipment and upgradation of our existing military hardware need assured allocation of funds. For the next decade or two, India's defence expenditure should be pegged at 3.5 per cent of the GDP, at constant prices, which will help us in inducting some modern weapons and equipment besides making up our large existing deficiencies. India's defence industrial base as well as its Research and Development capabilities are rather poor. During the last one decade or so, in the USA and Europe there has been a major consolidation in the defence industry with mega mergers of the global defence hardware manufacturers. Israel and South Africa have an advanced and sophisticated defence industrial base with reasonable access to Western defence technology. Besides Russia, India should enhance its defence cooperation with these countries to upgrade and modernise its defence industries. Opening up of defence production to civil sector would also help.

(c) While we do maintain a viable dissuasive posture against China, our offensive capabilities in mountains and high altitude areas are almost non-existent. Recycling and movement of troops against China from one sector to another is not an easy option. We must create separate land-air 'credible' counter offensive capabilities in the North and North-East for limited offensive operations into selected areas of Tibetan plateau. No war can be won on the basis of defensive deterrence however strong the defensive posture may be.

(d) India's major vulnerability is that it still does not have a missile force capable to effectively counter regional and extra regional threats. We need to attain IRBM capability which could reach Beijing and Shanghai even if launched from secure bases in central or southern India. We should also focus on building a credible nuclear force, mobile and widely dispersed that can survive first strike and yet be able to retaliate even if its command and control centres are wiped out. We need to strengthen our second-strike capabilities. Our nuclear deterrence should be 'robust' enough to deter any threat to our core interests in and around the subcontinent as well as in Indian Ocean. Notwithstanding the nuclear disparity with China, without proper nuclear deterrence capabilities, India will be totally naked to any Chinese threat. China's No-First-Use (NFU) is not applicable to India. As a part of our strategy to keep China constructively engaged, India and China should sign an unambiguous bilateral NFU Agreement.

(e) The increasing Chinese activities in the Indian Ocean and its periphery needs to be viewed seriously. An Indian naval capability including Rapid Reaction Amphibious force component needs to be created to tackle sea-borne threats to India's island territories and to sub-continental India. A strong Navy is a must for India to ensure the primacy of its interests in the Indian Ocean region.

Military to Military Contacts

The PLA enjoys considerable influence in Chinese Government and all their departments. To open another channel for communication between the two nations, it is in our interest to improve military to military relations with the PLA. We hardly 'understand' the PLA and its functioning. We need to interact to know more about them. The interaction between the two armies should initially be in the following fields and get intensified as we move along over a period of time.

(a) Reciprocal visits by the Service Chiefs and exchange of military delegations, heads of training institutions like National Dèfence College (NDC), College of Combat and the Indian Military Academy.

(b) Conduct of seminars on geo-strategic, economic and security related issues of global importance in general and of importance to Asia in particular, such as UN peacekeeping, regional economic co-operation, international terrorism and so on. An annual security dialogue involving India, China, Japan, Russia and the USA may be hosted by India.

(c) Exchange of officers for long courses including language courses, NDC and Staff College.

(d) Interaction in the fields of military, medicine and Defence Research and Development.

(e) Exchange of military delegations from premier training establishments.

(f) Sports and Adventure activities.

Management of Border Areas

India has long land borders with its six different neighbours i.e. 4096 km with Bangladesh, 3917 km with China, 3310 km with Pakistan, 1752 km with Nepal, 1458 km with Myanmar and 587 km with Bhutan. Borders depict the feeble, emotive, patriotic and nationalistic aspects of a country's psychological identity and physical security. As it is well proven in Asia, the border disputes can be trip wires to cross border tensions resulting in open hostilities. In the era of globalisation when the world is looking for secure and stable markets, India often figures in the world news as a working anarchy that seems to have a problem on almost all its borders. For example we have five disputed areas on our border with Bangladesh involving four different Indian states from West Bengal to Tripura. The terrain in our border areas varies from mountains to plains to jungles, marshy areas to riverine and is among the toughest to negotiate anywhere in the world. The disputed area in Siachen where we have on-going confrontation with Pakistan forms the world's highest battlefield. The border areas on our side are generally economically under-developed, sparsely populated and without any infrastructure worth the name. China has developed its border regions in a systematic manner over a period of time. Reportedly 80 per cent of the forward defence posts on the Tibetan plateau are connected with tarmac roads from the Chinese side. India has not earmarked sufficient resources for the development of its border areas. The tempo which was displayed after the Sino-Indian War of 1962 petered away with the passage of time, despite the fact that many parts of our border regions are believed to be rich in oil, gas and minerals. Special attention has to be paid towards infrastructure development. Emotional integration of North-East with rest of India is a sine-qua-non to preclude exploitation of its vulnerability by China.

Diplomatic Measures

To safeguard her national interests and to checkmate the increasing Chinese influence in its neighbourhood as well as in the Indian Ocean region, India needs to take some aggressive foreign policy initiatives. Defence and diplomacy are the two sides of the same coin for the attainment of national aims and objectives. There is a great need for India to further improve its relations with the countries of the region especially Vietnam, Myanmar, Thailand, Sri Lanka, Bangladesh, Indonesia, Malaysia and Singapore. Without logistics shore support, the possibility of Chinese Navy being able to operate in strength in the Indian Ocean till 2020 or so is rather remote. The scenario will change after 2050 in case China acquires a Blue Water Navy with a couple of aircraft carriers. This will also depend on number of other global developments which cannot be anticipated at this stage. We should make deliberate efforts to strengthen SAARC commercially despite our problems with Pakistan. Strategic and trade relations with Japan need to be boosted considerably. India's big market is an attraction for Japanese industry and investment provided the operating conditions are conducive and viable. We should also further develop our trade relations with Taiwan while politically and diplomatically maintaining a low profile. While building linkages at the political level with China and in different fields, India should prepare a fabric of interwoven relationship with different countries of the region. The image of India behaving like an 'elder brother' to its smaller neighbours must change. Cultivating friendly constituencies and web of relationships on our periphery will ensure our security at minimum costs. India should diplomatically dissuade Beijing from continuing its nuclear and missile aid to Pakistan by directly raising strategic costs for China. The costs could be increased through strategic cooperation with countries that share Indian concerns over China viz Vietnam, the Philippines and Japan. There is a good potential for military co-operation between India, Japan and ASEAN countries. The Central Asian Countries are traditionally mindful of China's growing military might. India can assist these countries in training and in servicing and upgradation of their indigenous military hardware. This cooperation is already there but its scope needs to be enchanced further.

The importance of the security of our sea-lanes can hardly be over-emphasised. An assessment of India's dependence on sea-borne trade for its very sustenance will reveal how crucial are the seas to India in the decades to come. A staggering proportion of India's foreign trade is sea-borne and accounts for as much as 97 per cent in terms of volume and 75 per cent by value. By the year 2003, 70 per cent of India's energy requirements will be met by imports via the sea and the figure is expected to grow exponentially over the next few decades unless fresh oil is struck in a major way in the Indian territory in the near future. Considering India's prolonged and troubled relations with its land neighbours, it is unlikely that its dependency over sea lanes for crucial requirements will recede in the decades to come. There is hence a need to vitalise a Bay of Bengal community for preserving the peace and security of this region. Further this community could well be a natural bridge between SAARC and ASEAN for exploiting maritime resources, trade and commerce as also encourage allied commercial activities including port development, tourism and cultural activities for the overall development and prosperity of the littoral states.

Sub Regional Cooperation

India and China need to be involved in sub-regional cooperative zone. India's North East and China's South-West could form a growth quadrangle with Myanmar and Bangladesh. The former two regions face similar developmental impediments such as being land-locked, economically backward, with a mountainous terrain and inadequate infrastructure. Both are blessed with a rich natural resource base. As political tensions between these countries further reduce, the 'do-it-alone' development policies will give way to cooperative framework. There is a vast scope for sub-regional fruitful cooperation in banking, railways, ship-building, internal water transportation, IT and power generation. These economic activities will in turn have cascading effect in further reducing tensions and boosting economic ties across national boundaries. Trade and investment ties will help foster a durable network of interests for which strong efforts should be made to broaden and intensify economic cooperation. Some measures which can be suggested are :

(a) Address the problems of information gap between the businesses communities or possible business opportunities for trade and investment.

(b) Provide direct airline and shipping links.

(c) Expediting and simplifying visa procedures.

(d) Sharing of information on several areas of mutual interest such as mega infrastructure projects, state sector reforms, labour issues etc.

India's North East, as BG Verghese says "is not just a distant extremity, as too long regarded, but a bridge to a most dynamic neighbourhood beyond with which it has always had close, historical, ethnic, cultural links." There should also be a sustained exchange and dialogue between academic and professional bodies. With such bilateral exchanges, an atmosphere conducive to building greater contacts and understanding can be built. Affiliations between related institutes could also play a vital role in this. Culture too offers a rich kaleidoscope of activities, which both countries would find mutually rewarding. In the area of sub regional cooperation between North East India and South West China, their strong Mongoloid content makes for several levels of cultural interaction. Cultural centres such as the North East Zonal Cultural Centre in Dimapur, which are involved in promotion of tribal and folk art forms, should be encouraged to interact and coordinate activities with respective cultural centres in China. There would also be several areas of common interest such as study of tribal languages, spiritualism, Tantrik studies, Buddhism and Tibetology. Holding of cultural festivals in each other's country would evoke interest among people by their increasing awareness.

Conclusion

It needs to be appreciated that at grassroot level there is considerable goodwill in China towards India, the land of the Buddha. Chinese generally have a good image of India as an ancient civilisation. Enhanced people to people contact will strengthen the bond of friendship between the two nations over a period of time which will minimise tension and reduce the risk of any hot or cold war between the two. India is a peace loving nation and will not do anything to aggravate the situation. In international relations there are no permanent friends or permanent enemies. On handling China we should analyse the growth and development of Sino-American relationship. Despite having fought a war and viewing each other as potential strategic rivals in the 21st Century, China and the USA are the biggest trading partners with extensive cooperation in various scientific, educational and cultural fields. We should constantly review our current and future environment to safeguard our national interests. We need to periodically review and update our overall military objectives based on our security perceptions. Such a culture is yet to find roots in India. The first ever Strategic Defence Review was undertaken in the year 1999. Now with a National Security Council in its place we should be able to streamline our security system much better than before. While we must continue to engage China at all levels in different fields, we should also keep the powder dry. Neglecting the development of our Armed Forces will cost us dear, as we had experienced in the Sino-Indian War of 1962. It is time for us to wake up and face the challenges ahead. Only an economically vibrant and militarily strong India will be able to face China in the years ahead.

Chapter 15

Conclusion

India's relationship with China during the later half of the previous century has covered vast and varied spectrum – the initial phase of amity and camaraderie, jolted by outright enmity followed by a protracted period of mutual distrust to a period of uncomfortable co-existence. The ensuing period has witnessed a new phase of constructive engagement and incremental co-operation between the two nations. By 2020 or so China will be a modernised regional military power and a potential economic super power. India is an aspirant for a permanent seat in a reorganised Security Council and will seek more prominent role in Asian affairs with a global reach. The Asian region today represents 60 per cent of the world's population, out of which India and China combined account for 40 per cent. Both the countries are representative of Asia's growing importance as the power structure shifts from Europe to Asia in the 21st Century. Both have an important responsibility to contribute to the establishment of a new international order for the development, well being and prosperity of the region.

The future of India-China relationship will depend upon international, regional and internal issues. The emerging Indo-US relations are likely to have an important bearing on India-China relations. At the regional level, competition over spheres of influence in Asia as well as clash of economic interests may create a conflictual situation. Stable and peaceful Sino-Indian relations can happen only if there is a general agreement on the terms of co-existence ; equality ; a fair settlement of territorial dispute and appreciating each other's security concerns as good neighbours. In order to sustain peace, India and China will need to arrive at a basic understanding of norms of political and strategic co-existence. This can only happen if India is economically and militarily strong. A strong India can only engage China on the basis of equality, dignity and reciprocity.

China is India's logical and even inevitable rival for regional influence and power. Clash of interests between the two is bound to take place–be it economic, sphere of influence or something else. It does not necessarily imply an open armed conflict between the two particularly in the near future, though the potential for such a conflict in the long run cannot be ruled out, be it along the disputed borders or high seas. The possibility increases as China's global stature and military capabilities increase. Asia has long seen a strong China and a weak India. There have been numerous occasions in history when China and India were simultaneously weak. But there has not been an instance when the two giants have been simultaneously economically and militarily powerful. That time is approaching fast. We, therefore, have to be ready for all eventualities, lest we are found wanting at the time of reckoning. In war there are no prizes for runners-up. We must engage China constructively. While we do this we must initiate various economic, diplomatic and military measures to safeguard our core values and national interests.

It must be stated that our understanding and knowledge about China, particularly about her armed forces is far from adequate. We do not know China well. Chinese society is in transition and under considerable transformation. It is essential that for a proper understanding of China, China watchers in India study the various developments intrinsically and draw their own conclusions rather than depend on the findings of the Western Sino observers. Such an expertise on China is yet to emerge in India. The sooner it comes about, the better.

Notes

Chapter 1: Introduction

1. June Teufel Dreyer, *China's Political System, Modernisation and Tradition*, London, Macmillan,2000.

Chapter 2: The Socio-Economic Developments

1. Colin Mackerras, *The New Cambridge Handbook of Contemporary China.*

Chapter 4: Changing Security Doctrines

1. Swaran Singh, *China's Changing National Security Doctrine*, New Delhi, IDSA, 1999.
2. Lonnie Henley, *PLA Logistics and Doctrine Reforms*, Pennsylvania, USA, Strategic Studies Institute, August 2000.

Chapter 5: PLA's Modernisation

1. James R Lilley and *China's Military Faces The Future*, Washington, American Enterprise Institute, 1999. David Shambaugh
2. S. Kondapalli, *The PLA: Evolving Dynamics*, New Delhi, IDSA, 1996.
3. S.Kondapalli, *China's Military: The PLA in Transition*, New Delhi, IDSA, 1999.
4. Paul Godwin, *Chinese Nuclear Forces: An Assessment* Cox's Report (Part II).
5. *White Paper* on China's National Defence in 2000
6. National Defence Report -Republic of China (Taiwan 1998).
7. US Strategic Studies Institute *The Chinese Armed Forces in the 21st Century*, December 1999.

8. US Strategic Studies Institute, *People's Liberation Army After Next.*
9. RAND Corporation , *People's Liberation Army in the Information ,\ge,* 1999.
10. Henry Stimson Centre, *"China's Foreign Military Relations"*, October 1999.
11. RAND Corporation Study, *Changing Political Military Environment,* April 2001.
12. June Teufel Dreyer, *China's Political System, Modernisation and Tradition,* London, Macmillan,2000.
13. *Defence & Foreign Affairs Handbook 2000,* International Strategic Studies Association.
14. "Long March to Modernization" *Jane's Defence Weekly,* 11 July 2001.
15. A Retrospective of Army Building, *News from China,* 01 August 2001.
16. Congressional Research Service, USA, *"China's Foreign Conventional Arms Acquisitions"*, October 2000.
17. 'Made in China', *India Today,* 25 June 2001.
18. *Chinese Military Digest,* 1999.
19. *China's Military : The PLA in 1990- 1991,* Boulder, West View Press, 1991.
20. FBIS-CHI-96-026 dated 07 February 1996, p.43.
21. *Jane's Defence Weekly,* 11 August 1999.
22. *JDW,* 15 September 1999.
23. *Jane's Armor and Artillery, 1998-1999* and *2000-2001.*
24. *,ane's Defence Weekly,* 27 October, 1999.

Chapter 6: PLA in Decision Making

1. Michael Swaine, *The Role of the Chinese Military in National Security Policy Making,* Washington, RAND, 1996.
2. Stimson Centre, *"China's Foreign Military Relations"*, October 1999.

Chapter 7: Chinese Intelligence System

1. *Chinese Intelligence Operations*, UK, Frank Cass & Co.
2. From the web: www.fas.org/irp/world/China

Chapter 9: Defence Industries

1. Richard Blitzinger, *China's Efforts to Leverage Advanced Technologies for Military Use.*
2. Cox's Report (Part II)
3. *Jane's Defence Weekly*, 23 September 1998.
4. *Jane's Fighting Ships*, 1999 -2000.
5. Pollack and Malvenon, *Assembled in China.*

Chapter 10: PLA's Logistics Support System

1. Lonnie Henley, *PLA Logistics and Doctrine Reforms*, Pennsylvania, USA, Strategic Studies Institute, August 2000.
2. "Striding Words: Modern and Great Logistics", *Foreign Broadcasts Information Service (FBIS)*, 06 August 1997.
3. "PLA Logistics Department", *FBIS*, 11 November 1994.
4. Tai Ming Cheng *Reforming the Dragon's Tail: Chinese Military Logistics in the Era of High Technology Warfare and Market Economies.*

Chapter 11: China's Activities in India's Neighbourhood

1. RAND Corporation, *The Changing Political- Military Environment in South Asia*, (Washington DC , April 2001).
2. *News Behind News*, 16 July, 30 July and 13 August 2001.
3. Vijay Kranti, "Chequered History", *The Pioneer*, 19 July 2001.
4. " Burma Cats Wary Eye on China", *Jane's Intelligence Review*, June 1999.
5. SC Bhatt, *The Triangle: India, Nepal, China*, New Delhi, Gyan Publishing House,.
6. Gautam Sen,"Indian Foreign Policy Today", *Sanghsandesh*, Jan - February 2001.
7. G Parthasarathy, "Errant RAND Gets Real About India", *The Pioneer*, 16 August 2001.

8. AK Sen, "Chinese Firm Supplies Missiles to Pak: Violates Pledge", *The Asian Age,* 08 August 2001.
9. "Asian Security and Role of Outside Powers in the Region", *Proceedings of Joint USI-CIISS (China) Seminar* held at USI, New Delhi (March 2001).
10. Lt. Cen. Chandra Shekhar, "Security Perspective and Geo Strategic Realities in the Indian Context", *USI Journal,* July-September 2001.

Chapter 12: Development in Tibet

1. *Tibet 2000,* Central Tibetan Administration, Dharamshala.
2. *Study on Deployment of Chinese Forces in Tibet-* by Tashi Chutter.
3. *Demilitarization of Tibetan Plateau,* Central Tibetan Administration, Dharamshala.
4. (Tibet) *Behind Bars,* New Delhi, Indraprastha Press.
5. *Tibet Information Network News Update,* 13 June 2001.
6. *Report on Fourth Tibet Work Forum,* TIN News Update, 27 July 2001.
7. General K. V. Krishna Rao, *Prepare or Perish: A Study of National Security,* New Delhi, Lancer, 1991.

Chapter 13: China as a Strategic Challenge

1. Vinod Saighal, Maj Gen, "Dealing with China in the Twenty First Century", *USI Journal,* January -March 1999.
2. Swaran Singh, "Building Security and Confidence with China" in *Across the Himalayan Gap.*
3. Ma Jiali, "Striving to Establish a Constructive Cooperative Partnership between India and China", Extract from *China Report,* 36 : 3, 2000.
4. Raghavan V R, Lt Gen, "India-China Relations: A Military Perspective" in *Across the Himalayan Gap.*
5. Sreedhar, "China Becoming a Super Power and India's Options", in *Across the Himalayan Gap.*

6. Malik Mohan, *The India -China Divide.*
7. Ramesh Thakur, *Normalizing Sino-Indian Relations.*
8. Swaran Singh, "Sino-South Asian Ties", *Strategic Analysis,* April 2000.
9. Ashley Tellis, *Stability in South Asia.*
10. Ashley Tellis, *China's Grand Strategy.*
11. Study Tour to Boost Sino-Indian Trade in NCR, *The Pioneer,* 06 September 2001.
12. General K. V. Krishna Rao, *Prepare or Perish: A Study of National Security,* New Delhi, Lancer, 1991.
13. Samuel S Kim, *China's Quest for Security in the post Cold War World,* Boulder, Westview, 1999.

Bibliography

Primary Sources

Congressional Research Service, USA, *"China's Foreign Conventional Arms Acquisitions"*, October 2000.

Defence & Foreign Affairs Handbook 2000, International Strategic Studies Association.

Demilitarization of Tibetan Plateau, Central Tibetan Administration, Dharamshala.

Godwin, Paul, *Chinese Nuclear Forces: An Assessment*.

Cox's Report (Part II)

Military Balance2002 – 2003 (UK, IISS, 2003).

National Defence Report-Republic of China (Taiwan 1998).

The New Cambridge Handbook of Contemporary China by Colin Mackerras.

Report on *Fourth Tibet Work Forum* (TIN News Update dated 27 July 2001).

Tibet 2000, Central Tibetan Administration, Dharamshala.

White Paper on China's National Defence in 2000.

Secondary Sources

Bhatt SC, *The Triangle: India, Nepal, China,* New Delhi, Gyan Publishing House,.

Blitzinger, Richard, *China's Efforts to Leverage Advanced Technologies for Military Use.*

Dreyer, June Teufel, *China's Political System, Modernisation and Tradition,* London, Macmillan,2000.

Garner Robert E, *Understanding Contemporary China,* US, Lynne Rienner, 1999

Henley, Lonnie,	*PLA Logistics and Doctrine Reforms*, Pennsylvania, USA, Strategic Studies Institute, August 2000.
Kim, Samuel S,	*China and the World : New Directions in Chinese Foreign Relations*, Boulder, Westview, 1999/ *China's Quest for Security in the post Cold War World.*
Kondapalli, S,	*The PLA: Evolving Dynamics*, New Delhi, IDSA, 1996.
Kondapalli, S, David Shambaugh,	*China's Military: The PLA in Transition*, New Delhi, IDSA, 1999.
Lilley, James R and	*China's Military Faces The Future*, Washington, American Enterprise Institute, 1999.
Raghavan V R, Lt Gen,	"India-China Relations: A Military Perspective" in*Across the Himalayan Gap.*
Rao, K. V. Krishna General	*Prepare or Perish: A Study of National Security,* New Delhi, Lancer, 1991.
RAND Corporation,	*People's Liberation Army in the Information Age,* 1999.
RAND Corporation,	*The Changing Political-Military Environment in South Asia*, (Washington DC , April 2001).
Swaran, Singh	*China's Changing National Security Doctrine*, New Delhi, IDSA, 1999.
Swaran, Singh	"Building Security and Confidence with China" in *Across the Himalayan Gap.*
Sreedhar	"China Becoming a Super Power and India's Options", in *Across the Himalayan Gap.*
Stimson Centre,	"*China's Foreign Military Relations*", October 1999.
Tai Ming Cheng	*Reforming the Dragon's Tail: Chinese Military Logistics in the Era of High Technology Warfare and Market Economies.*
Tellis Ashley,	"*China's Grand Strategy*".
Thakur, Ramesh,	*Normalizing Sino-Indian Relations.*
Swaine Michael,	*The Role of the Chinese Military in National Security Policy Making*, Washington, RAND, 1996.

US Strategic Studies Institute, *The Chinese Armed Forces in the 21st Century,* December 1999.

US Strategic Studies Institute, *People's Liberation Army After Next.*

Journals/Newspapers/Web

"Asian Security and Role of Outside Powers in the Region", *Proceedings of Joint USI-CIISS (China) Seminar* held at USI, New Delhi (March 2001).

"Burma Casts Wary Eye on China", *Jane's Intelligence Review,* June 1999.

"Chinese Firm Supplies Missiles to Pak: Violates Pledge", AK Sen, *The Asian Age,* 08 August 2001.

Chinese Military Digest, 1999.

"Errant RAND Gets Real About India", C Parthasarathy, *The Pioneer,* 16 August 2001.

"Indian Foreign Policy Today", Gautam Sen, *Sanghsandesh,* Jan -February 2001.

Internet: www.fas.org/irp/world/China

Jane's Fighting Ships, 1999 -2000.

Jane's Defence Weekly, 27 October, 1999.

"Long March to Modernization" *Jane's Defence Weekly,* 11 July 2001.

'Made in China', *India Today,* 25 June 2001.

A Retrospective of Army Building, *News from China,* 01 August 2001.

Saighal, Vinod, Maj Gen, "Dealing with China in the Twenty First Century", *USI Journal,* January -March 1999.

"Security Perspective and Geo Strategic Realities in the Indian Context", Lt. Gen. Chandra Shekhar, *USI Journal* July-September 2001

"Sino-South Asian Ties", Swaran Singh, *Strategic Analysis,* April 2000.

"Striving to Establish a Constructive Cooperative Partnership between India and China", Ma Jiali, Extract from *China Report,* 36 : 3, 2000.

Index